The Destruction of Sodom

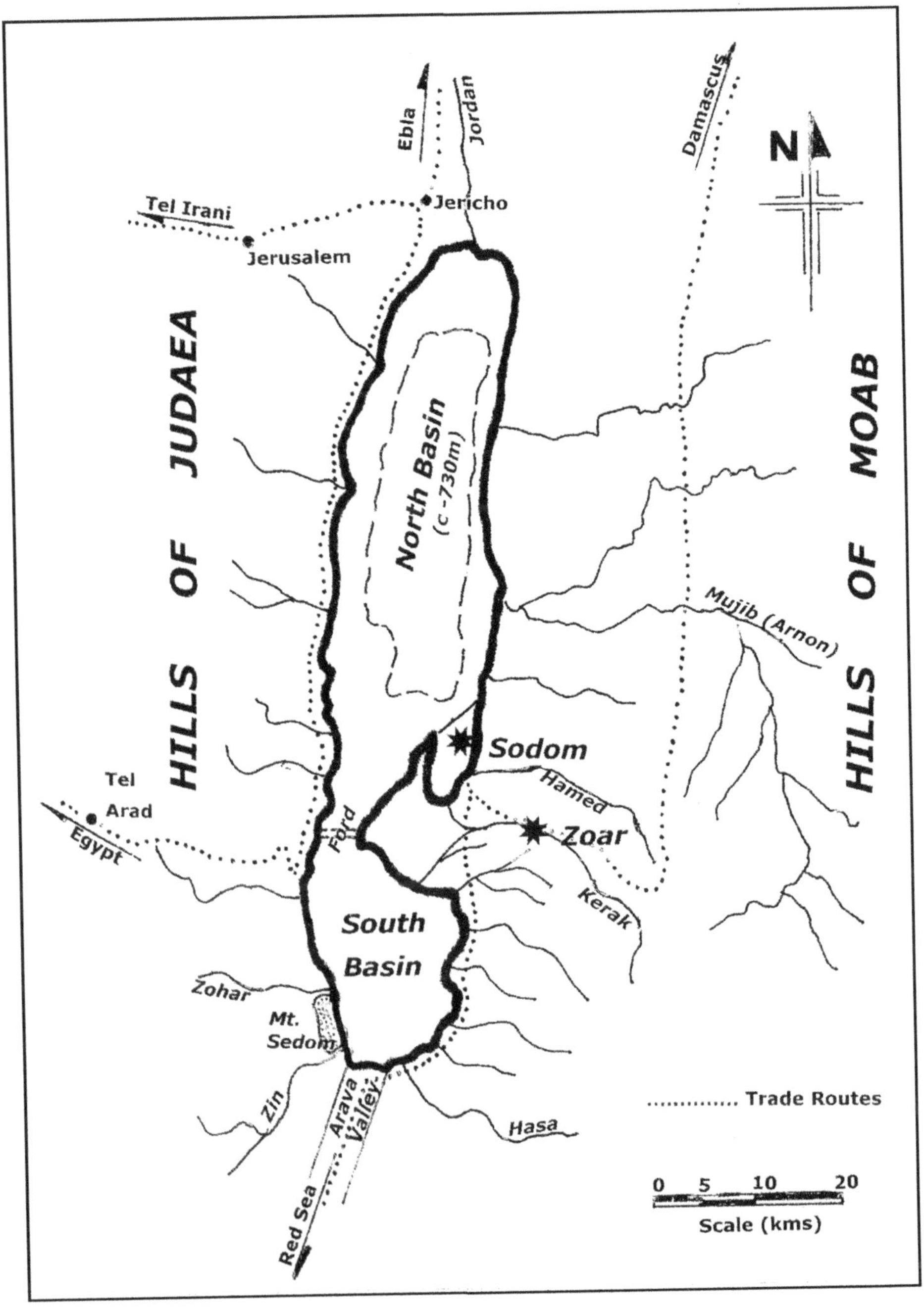

Map of the Dead Sea,
showing the postulated location of Sodom and Zoar.

The Destruction of Sodom

A Scientific Commentary

Graham Harris

The Lutterworth Press

The Lutterworth Press
P.O. Box 60
Cambridge
CB1 2NT
United Kingdom

www.lutterworth.com
publishing@lutterworth.com

ISBN: 978 0 7188 9368 2

British Library Cataloguing in Publication Data
A record is available from the British Library

First Published by The Lutterworth Press, 2015

Contents

List of Figures

List of Tables

Preface

This volume had its origin many years ago after a television documentary was aired with the purpose of locating the remains of Sodom and Gomorrah, those legendary cities in the Vale of Siddim so vividly destroyed by the wrath of God. The documentary was of above average interest to me, as I had spent the best part of a decade involved in a number of geological studies in the Dead Sea valley, where Sodom, Gomorrah, and their satellite cities, were popularly located.

In the documentary, a number of scholars advanced theories of where they thought Sodom (or the other cities mentioned in Genesis) may or may not have been. Members of one profession were notable by their absence from this gathering – geologists! A need was thus perceived for a publication describing the geology of the Dead Sea written for archaeologists, in the hope it might prove useful to this age-old search. A colleague and I set about the task of collecting all information of possible relevance, which included plundering my personal files for data relating to the properties of the soft sedimentary deposits that had accumulated within the Dead Sea valley.

During this filtering process it began to dawn upon us exactly where the Vale of Siddim had been located, and in what fashion Sodom, Gomorrah, and their satellite cities, had been destroyed so dramatically. It also became apparent that the region had been of considerable importance to the commerce between Canaan and Egypt. The loss of the cities in such an inexplicable manner would have devastated this trade.

The results of our research were published in *The Quarterly Journal of Engineering Geology* (November 1995), a publication of The Geological Society, London, England. The paper was titled 'The Destruction of Sodom and Gomorrah: a geotechnical perspective', and was well received by the technical readership to whom it was addressed. The paper came to the attention of the BBC, London, while planning their series *Ancient Apocalypses*, and was utilised in their television documentary dealing with the destruction of Sodom and Gomorrah, first aired in 2001. During the preparation of the documentary the BBC commissioned laboratory testing at the Schofield Centre for Centrifuge Studies, Cambridge University, which validated the postulated mode of destruction; namely soil liquefaction by earthquake.

Since limitations of space in the one, and time in the other, prevented more comprehensive treatment in either, it was thought useful to compile the present volume in order to remedy this deficiency. Furthermore, new information has since been acquired which is included herein. The objective of this work is to draw together the related strands of knowledge from a spectrum of scientific and cultural disciplines, in order to give greater clarity and deeper understanding to a noteworthy event which impacted the region during the third millennium, and which is so vividly described in the Book of Genesis. The destruction of Sodom, and its satellite cities, is concluded as having led to political instability in Egypt at the end of Early Bronze III. Consequently it is hoped this contribution will be of interest, not only to scholars and spur further research into this singular event, but also to the general public.

A critical reviewer for *Archaeology* magazine stated quite bluntly with regard to the story of Sodom and Gomorrah, that 'The real challenge for biblical archaeologists today is not to search for long-lost cities, but to understand why the ancient Israelites formulated these powerful myths' (Silberman, 1996). This dismissive attitude may reflect the opinion of some scholars and archaeologists, but it is irrational. The search for the remains of any city of Antiquity, whether Biblical or not, is a source of fascination to a very wide public, and more enduring than most if the city concerned possesses the legendary notoriety of Sodom and Gomorrah. Indeed, though the destruction of the 'cities of the plain' may have its origins in myth, a responsible attitude of any scholar is to analyse that myth rationally and scientifically with all the tools available. We are right to ask questions, and advance hypotheses, provided we take into account all evidence from whatever source it may originate. Under the present circumstances, as it relates to the destruction of Sodom and Gomorrah, the most effective research is multi-disciplinary.

Here the approach has been to place the narrative against the backdrop of all relevant established geographic and scientific fact. Bluntly stated it is 'to put the horse before the cart', not the reverse. The legendary demise of Sodom and Gomorrah, long buried in the mists of antiquity, embedded in oral tradition, and intertwined with moralistic teaching, likely constitute kernels of historic truth, but they cannot be corroborated by cultural considerations alone no matter how intense the scholarly debate. To make any credible attempt to unravel the truth requires applying the powerful tools of scientific knowledge now in our possession. Therefore, the framework of this book places great emphasis on various branches of the earth sciences, such as geography, geology, seismology, geotechnology, and petroleum chemistry, set against an abbreviated backdrop of archaeology, history, and religion.

One of the great pioneering archaeologists in the search for Sodom was Dr Melvin Kyle. In his famous book *Explorations at Sodom* (1928) he writes in his Introduction 'I conceive the popular telling of dry-as-dust facts on a recondite subject to be one of the greatest services a writer may render.' The author echoes the sentiments of Dr Kyle in the hope that the translation of 'dry-as-dust facts', so necessary for the argument presented herein, may be accomplished with a minimum amount of tedium. In compiling this work I have made every effort to avoid arcane or complicated language, however, by necessity scientific terms are unavoidable, and some forbearance is requested by the reader should these prove unfamiliar. To this end, a glossary has been included as an aid to understanding some of the more specific terminology.

A publication such as this necessarily includes information extracted from a wide variety of sources compiled by others, and it goes without saying that a great debt is owed to all those whose works are referenced herein, and to whom acknowledgement is duly made. The placement of photographs in the public domain by both individuals and government agencies, especially the United States Geological Survey, has greatly facilitated the illustration of this monologue.

Special mention must be made of the work of Kenneth Emery and David Neev, whose extensive research into the geology of the Dead Sea, and the chemistry of its waters, established a solid factual basis for all subsequent work.

Tribute must be made also to Professor Harry Bolton Seed of the University of California, Berkeley, who gave so readily of his specialist knowledge in earthquake engineering and encouraged this work, and to Professor Nicolas Ambraseys of Imperial College, London, under whom the writer was privileged to study for a short period. It is to be regretted both passed away before the thesis presented herein could be finalised. Appreciation is also extended to Dr John Cripps, of Sheffield University, who encouraged this publication.

Information in my own possession arose during involvement with extensions to the Dead Sea Works Limited chemical plant at S'dom, Israel, while in the employ of Bateman Engineering Limited, Johannesburg, South Africa. Gratitude is accordingly expressed to both of these organisations. Also, I would like to publicly acknowledge the valuable assistance rendered by Tony Beardow during the writing of our original joint paper, the incentive provided by the BBC through inviting me to participate in their documentary, and to Dr Jonathan Tubb of the British Museum. It was at Jericho where I first met Dr Tubb, to discover both of us had attended the same school – King Henry VIII Grammar School, Coventry, England. That evening we recalled some of the teachers we shared those years of long ago. Perhaps, in some small way, they too have contributed to this volume.

Acknowledgements

In addition to Graham's acknowledgements made in the preface of this book, the following additions are made:

A debt of gratitude is expressed to the editorial staff of Lutterworth Press, who agreed to continue with the publication of this book after Graham's sudden death in July 2014. It is hoped that this will be a true legacy for his efforts in compiling this work.

Special thanks are also expressed to Les MacPhie, a long-time friend and former colleague of Graham, for stepping into the breach and assisting the editorial staff in the copy-editing of this book. His help has been invaluable.

Susan Harris

Introduction

The narrative of the destruction of Sodom and Gomorrah and the transformation of Lot's wife into a pillar of salt, as related in the Book of Genesis, has intrigued generations. Destroyed as a consequence of God's displeasure with the inhabitants because of their lax morals, the five 'cities of the plain' (Sodom, Gomorrah, Admah, Zeboiim, and Zoar) have presented archaeologists with both an enigma and a challenge. Did the cities exist? If so, where? Answers to these questions have hitherto defied systematic search. The two issues remain unresolved.

There are few who haven't heard the story of how God destroyed these ancient cities with fire and brimstone, saving only Lot and his daughters. Genesis depicts Lot's wife being turned into a pillar of salt because she looked back at the scene of carnage from which they were fleeing. St. Luke's gospel (17:32) warns 'Remember Lot's wife', and St. Peter in his second epistle (2:6-7) also makes reference to the destruction of Sodom and Gomorrah, as does Ezekiel (16:49-51). Over the centuries numerous sermons and homilies have been delivered using this dramatic Old Testament story, urging people to turn from sinful ways into the paths of righteousness.

How real was the drama told in this tragic and spectacular tale? When a violent and vivid event like this occurs in the very dawn of history, an event which cannot be rationalised or understood at the time, it is inevitable that myth and legend arise and, as a consequence, the disaster is assigned to immorality and a wrathful God.

The dictionary defines myth as a 'fable containing an historic or moral truth', whereas a legend is 'an unauthenticated fable'. This book might well be subtitled 'Post-Mortem on Sodom', being an attempt to dismiss both myth and legend surrounding the Biblical account, and to rationalise the destruction using scientific knowledge and argument. This does not displace any theological perspective, or even dispute the account in the Book of Genesis. As will be seen it augments these by presenting a solid technical argument which accounts not only for the existence of Sodom, but the mechanics of its catastrophic demise. It is hoped it will prove useful to those interested in the resolution of this ancient enigma, whatever their profession, background, or religion.

Numerous scholars have concluded from the evidence available that the legendary cities did exist, even though they may dispute their locations:

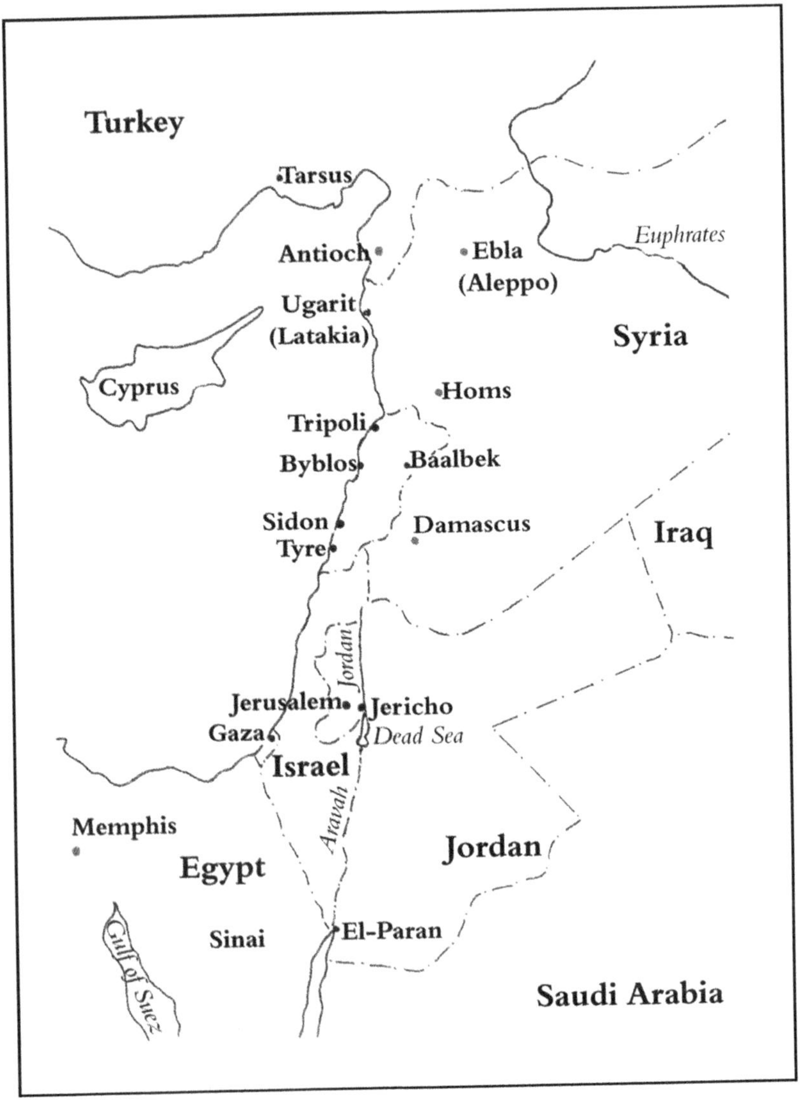

1. Regional map of the Near East.

There can hardly be any doubt that the Biblical traditions about Sodom and Gomorrah . . . are only part of a complex of ancient and hazy recollections that go back to some actual event. (Vawter, 1977)

It is quite possible that the tradition contains a distant recollection of an ancient catastrophe. (Rad, 1971)

In 1975 a great archive of clay tablets dating to 2400-2350 BC was discovered at Tell Mardikh, ancient Ebla, in northern Syria. One of the tablets is a geographic atlas listing 289 place names. An analysis of two

> segments of the list by Shea (1979) indicates that they are sites located in Palestine, possibly places visited by merchants. The second segment traces a route from Syria south through the central hill country of Cisjordan, along the western shore of the Dead Sea, south of the Dead Sea Plain and then north along the east side of the Plain and Dead Sea. In the area corresponding to the east side of the Dead Sea Plain there are two places named – Admah and Sodom. (Wood, 1999)

> The unearthing of the royal archives at Ebla in 1976 has brought many new facts to light. It was reported then denied, that one of the Ebla tablets lists the five 'Cities of the Plain' in the same order as Genesis (Ch 14). (Hattem, 1981)

The discovery of the Ebla tablets at Tell Mardikh in 1975, near Aleppo, Syria, was a watershed in Middle Eastern archaeology. Almost fifty years earlier, in 1929, a similar discovery had unearthed Canaanite tablets on the Mediterranean coast at Ugarit, now Ras Shamra, near Latakia, Syria. The initial find at Ebla resulted in the exhumation of 1800 intact tablets, 4700 in fragments, and numerous chips relating to the period c. 2500-2250 BC after which the city was destroyed. More tablets were discovered in later digs, the total understood to number between 4000 and 5000. Though the clay tablets had not been fully translated it was clear that by 2500 BC Ebla was politically influential, and possibly dominant, over a vast region extending as far south as Damascus.

Dr Giovanni Pettinato, a Sumerologist who was engaged in the deciphering of the tablets, noted they possessed a markedly Sumerian influence, and concluded the Sumerian writing system was used by the scribes as a basis of composing a non-Sumerian West-Semitic text, i.e. Syro-Palestinian or Caananite.

It was apparent that the Eblaites worshipped a pantheon of gods, some five hundred all told. However, there was a hint of henotheism, i.e. the worship of a supreme god above all other gods. There were also lists of geographical place names with which the king of Ebla conducted trade, and among these Pettinato reported (in 1976) that all five 'cities of the plain' were mentioned, but in reverse order to those listed in Genesis. He was later to revise his statement to the effect that only the cities of Sodom and Gomorrah could be positively identified, but their names were found on more than one tablet.

The importance of the discovery of the tablets at Ebla cannot be underestimated. Not only did they reveal Ebla to have been a capital city, but one which was large and powerful in terms of trade and commerce, with wide-ranging influence. According to Roux (1964) the tablets provide a wealth of information on the organisation, social structure, economic systems, diplomatic and commercial relations, areas of influence, and cultural affinities of this long forgotten kingdom, and will continue to do so as long as studies continue into their contents. No city-state of the period has bequeathed such a legacy.

Professor Biggs (1974, 2005) of the University of Chicago, published a list of geographical place names discovered at Abu Salabikh (ancient name unknown), close to Nippur, the capital of ancient Sumer. This list received scant attention the first time it was published, as it was compiled in Sumerian logograms and, thus, largely unintelligible. However, the list from Ebla was spelt phonetically, thereby facilitating direct comparison of the two discoveries.

The mention of Sodom and Gomorrah on the Ebla tablets implies two things. Firstly, the cities existed at the time of their writing. Secondly, that trade connections existed between Ebla and the 'cities of the plain'.

Though pictograms were used in cuneiform texts, and hieroglyphs in Egyptian texts, to denote objects and numbers, written language appears to have evolved not much earlier than about 1000 BC. The first five books of the Old Testament, the Pentateuch, are considered to have been compiled in the present format mostly in the period 400-300 BC. The vivid nature of the catastrophe that engulfed Sodom and Gomorrah would have ensured its inclusion in the first book of the Pentateuch, i.e. Genesis, if only for the moral lessons the event demanded; prior to that an oral tradition would have prevailed.

The destruction of the temple in Jerusalem by the Romans in AD 70 resulted in the loss of many ancient Hebrew scrolls. This entailed much rewriting. It would have been perfectly natural for any scribe, in the predicament he found himself, to have drawn upon all available sources to resurrect Hebrew history, and this would not have excluded recourse to myth, legend and folklore. Perhaps, as a consequence, it is inevitable that a dichotomy relates to certain passages, in particular to the story of Abraham and Lot set against the dramatic backdrop of the destruction of Sodom.

Though there is consensus of the actual existence of Sodom and Gomorrah in the vicinity of the Dead Sea, there is little consensus among Biblical scholars regarding how or when the cities met their fate. Hattem (1981) quotes Freedman (1977) as saying:

> While I do not think we can pin down the dates of the Genesis 14 stories precisely, it seems to me most likely that they belong to the period around 2400 to 2100 BC. Attempts by other scholars to resolve this issue more closely, with no input of science, have generally relied on genealogical studies of the Patriarchs – Abraham, Isaac, Jacob and Joseph.

Tubb (1998) gives his opinion from an archaeological viewpoint, stating:

> The question of the historicity of the Patriarchal narratives is a complicated and largely unanswerable one. Archaeology cannot hope to prove the existence of Abraham or Isaac, or indeed any of the other characters referred to in the book of Genesis. On the other hand there is little reason to doubt that in very general and very basic terms the founding fathers of what was to become Israel might well have entered Canaan at some stage

> during the third or early second millennium. Unfortunately this is about all that can be said. For it must be appreciated that the stories relating to the Patriarchs were not put into their final form until sometime between the seventh and fifth centuries BC.

Alter and Kermode (1987) stated bluntly – '[The Old Testament] is a rather loose anthology that reflects as much as nine centuries of Hebrew literary activity.' This suggests a certain degree of flexibility is desirable in interpreting Biblical texts.

Scholarly opinion suggests the likelihood that the writers of Genesis, preoccupied in recording the Patriarchal narrative, included the Sodom and Gomorrah event to demonstrate God's power for a multitude of possible reasons. The power, which had manifested itself in the Creation of Earth and all therein, could also be summoned to wipe out those who lived in sin or disobedience. A dire warning if ever there was!

Trifonov (2007) expressed the opinion that the destruction of Sodom and Gomorrah with brimstone and fire, as depicted in the Bible, indicated influence of volcanic action. He suggests the fate of Sodom and Gomorrah more likely rested with earthquake and flood as the Dead Sea basin shows no evidence of recent volcanism. This suggested to him that the oral tradition from these two separate geologic events may have been merged in the Biblical account. The closest location to the Dead Sea valley where volcanic evidence has been noted is at two settlements in southern Syria, Khirbet El-Umbashi and Hebariye, where many animal bones were discovered buried in basaltic lava. The settlements were dated to the latter half of the third millennium BC. He concludes:

> Located in the Dead Sea region, Sodom and Gomorrah were most probably destroyed by a strong earthquake or flood [or both], but the fresh memory about two settlements perishing from a volcanic eruption caused the population to merge these two events.

Viewing the incident through twenty-first century eyes, it is tempting to ask what other cities have been similarly destroyed since in like manner, for equally just cause. The moralist would have little difficulty in assembling a long list of potential candidates for destruction, both in the past and in the present – but God has withheld His hand and, apparently, continues so to do. His wrath does not appear to have been repeated in such a dramatic way, as we are urged to believe was inflicted upon the inhabitants of Sodom and Gomorrah. Why? It is difficult to accept that the citizens of Sodom were any more sinful in their sexual perversions than the world today! And if, as is concluded in this book, the cities of Sodom and Gomorrah had a sound economy based on the industrial production of a desirable commodity, which was tantamount to a virtual monopoly, then the behaviour of its citizens is likely to be no better, or any worse, than the average industrial centre of the modern era fortunate to be bequeathed with similar riches.

The conclusion must be that the destruction of Sodom and Gomorrah was not the unique moral nemesis as portrayed by the writers of Genesis. The fact it can be explained by the interaction of natural physical processes goes a long way in providing reassurance that the God of Creation is, indeed, a long-suffering and merciful God. Nevertheless, it is understandable that the mysterious circumstances of the destruction of those notorious cities supplied fuel for much priestly speculation in a bygone age when superstition ruled on matters inexplicable.

One of the difficulties faced by archaeologists in dating events of the long distant past, whether those events are Biblical or not, is that for centuries life changed very little or, when it changed it changed slowly. By contrast the present age is one where changes of lifestyle often occur on a daily basis. To choose a trivial example, a person's age can be determined fairly accurately by which film star they remember best, or what music recordings they cherish whether on vinyl, tape or CD. But four to five millennia ago the sameness of everyday life understandably presents hurdles in dating events with any great accuracy. In tackling this problem archaeologists have placed great reliance on the identification of pottery shards, among other artefacts, as changes in pottery manufacture, glazing and decoration during the course of history helps to establish a timeline of events. It also assists in determining patterns of trade.

Archaeological investigations in the vicinity of the Dead Sea have identified a number of sites which can be traced back to the Early Bronze and Middle Bronze ages. The most notable of these are Safi, Numeira and Bab edh-Dhra located towards the southern end of the Dead Sea. It is tempting to associate some of these with the infamous 'cities of the plain'. However, such speculation is unprofitable if we are to place any reliance on the Genesis account in our search for Sodom, as the account records in vivid detail how the cities were totally obliterated, suddenly and without trace, with the exception of Zoar. It was a perfect apocalypse. Therefore, it is obvious that if there aren't any remains of Sodom or Gomorrah (at least on land) then, *a priori*, neither Safi, Numeira, nor Bab edh-Dhra can be considered prime candidates for either. Also, placing five of the reputedly most degenerate cities of antiquity on a salt plain in the South Basin of the Dead Sea, as some have suggested, hardly makes sense. From the aspect of town planning alone, and the necessity of possessing a reliable source of fresh water for domestic consumption, it is obvious the locations of Sodom and Gomorrah, and their satellite cities, must be sought on the alluvial fans, of which there are several in the general area.

The Greek geographer Strabo (c. 54 BC – AD 21) travelled extensively throughout the Roman world in the compilation of his monumental work *Geographica*. In this he describes the Dead Sea in great detail, and provides a useful commentary regarding the destruction of Sodom and Gomorrah. Strabo is reported as being well acquainted with Egypt, so it is likely he visited the Dead Sea. *The New Encyclopaedia Britannica* (fifteenth edition, 1992) states that:

> Strabo's descriptions of Mesopotamia, Syria, Palestine, and the Red Sea (Book XVI) were based on the accounts of the expeditions sent out by Mark Antony (about 83 to 30 BC) and by the emperor Augustus, as well as on chapters on ethnography in Posidonius and on the book of a Red Sea voyage taken by the Greek historian and geographer Agatharchides (second century BC).

Bunbury (1959) says this of Strabo's work:

> The *Geography* of Strabo is not only the most important geographical work that has come down to us from antiquity: but it is unquestionably one of the most important ever produced by any Greek or Roman writer. It was indeed, as far as we know, the first attempt to bring together all the geographical knowledge that was attainable in his day, and to compose what would be called in modern times a general treatise on geography.

One of Strabo's descriptions of the Dead Sea follows:

> Many other evidences are produced to show that the country is fiery; for near Mossada are to be seen rugged rocks that have been scorched, as also, in many places, fissures and ashy soil, and drops of pitch dripping from smooth cliffs, and boiling rivers that emit foul odours to a great distance, and ruined settlements here and there; and therefore people believe the oft-repeated assertions of the local inhabitants, that there were once thirteen inhabited cities in that region of which Sodom was the metropolis, but that a circuit of about sixty stadia [11 kilometres] of that city escaped unharmed; and that by reason of earthquakes and eruptions of fire and of hot waters containing asphalt and sulphur, the lake burst its bounds, and rocks were enveloped with fire; and as for the cities, some were swallowed up and others abandoned by such as were able to escape. (Book XVI, 2, 44)

This description by Strabo is of considerable value, because it was written by a geographer well versed in recording factual observations with the greatest accuracy possible, whether or not they were his own. Though written some two thousand years after the event it is, nevertheless, of inestimable value. It is also the only written evidence we have other than that recorded by the writers of Genesis, thereby giving credence to the Biblical account.

If, as Strabo states, 'Sodom was the metropolis' then it follows there should be some economic *raison d'être* for the existence of Sodom and its satellite cities. This can be explained, in part at least, by the exploitation of the region's bitumen resources for which the Dead Sea was renowned. The Romans named the Dead Sea *Lacus Asphaltitis*, and Strabo describes how the bitumen was harvested from the sea when risings occurred, thereby indicating bitumen production to be still practised during the Roman occupation. Therefore, the bitumen industry of the

Dead Sea can be concluded as having not only existed for many centuries, but was of considerable economic importance both before and after the destruction of Sodom.

It is likely that this industry was supported by the extraction of other useful raw materials, e.g. salt, limestone, and silica sand, which abound in the region. Therefore, it is not beyond the bounds of possibility that the Sodomites represented the world's first oil refiners and chemical engineers, crude though their engineering talents may have been in comparison to ours today.

Assuming Sodom to be the prime centre of commercial activity, one must focus on the area where bitumen risings predominate for its location, and where an abundant supply of fresh water is assured to support a large metropolitan population, probably the largest population of any of the 'cities of the plain'.

Historical studies into the seismicity of the Dead Sea valley have concluded that the destruction of Sodom can be dated to about 2350 BC, whereas Biblical scholars generally attribute the story of Abraham and Lot to some three to four centuries later, i.e. 1900-1800 BC, though there is no particular consensus. This seeming contradiction does not discredit the Biblical account of the destruction of Sodom, but instead it enhances it. By its selection as a backdrop to the Abraham-Lot saga, which is central to much of Genesis, it adds considerable force to the moral message.

Neev and Emery (1995) go further in linking the destruction of Sodom and Gomorrah to a cataclysmic seismic event. They write:

> The destructive potential of earthquake energy affecting the trough of a basin filled with uncompacted sediments [i.e. the Dead Sea] would be much greater than along its solid perimeter, as illustrated by the 20th-century earthquakes at Mexico City, San Francisco, and Los Angeles. The Sodom and Gomorrah event could have been even more severe because other factors were involved in it, such as the gliding of sedimentary fill [i.e. a landslide] toward the centre of the basin. . . . Biblical script derived from oral tradition of the disruption of Sodom and Gomorrah as a supernatural event should not be dismissed as an exaggerated fictitious story written to satisfy religious motives.

By these remarks Neev and Emery do not rule out a liquefaction induced landslide which led to the destruction of Sodom and her satellite cities, followed by an equally devastating tsunami. In fact they imply its every possibility.

As a people, what were the inhabitants of Sodom like? The Biblical account portrays the Sodomites as lewd and lustful, prosperous but profligate. This view may or may not be true, but the moral tone of Genesis suggests it may be a sweeping generality which, of course, is incapable of authentication.

Rappoport in his *Ancient Israel: Myths and Legends* (1928) retells many of the familiar Bible stories, but he draws upon a wider selection of sources claiming to have consulted original documents in Hebrew, Aramaic and Greek, the majority of which are inaccessible to the average reader. One of these appears to be the apocryphal Book of Jasher, which though of doubtful authenticity is considered to possibly represent a 'lost book of the Bible', as various stories are duplicated in the Old Testament. In one chapter – The Sin of Sodom – Rappoport depicts the Sodomites as mean and grasping, as does the Book of Jasher. He writes in the following terms:

> Moreover, the inhabitants of Sodom never gave of their substance to the poor and the alien who came to their city. They passed a law according to which all aliens were to be expelled, and poor men seeking food never to receive a piece of bread. Charity was a crime punished heavily within the walls of Sodom. The men of Sodom elected as their judges and rulers men of falsehood and wickedness, who mocked justice and equity and committed evil deeds. Whenever a stranger happened to enter the city of Sodom, the inhabitants at once took away his goods and substance, divested him of his clothes, and sent him away poor and naked. There was no use to appeal to the laws of the city, for the judges themselves approved such deeds.
>
> Thus the inhabitants of Sodom, who knew neither charity nor human kindness, waxed rich and exceedingly prosperous, and lived in peace. They even passed a law that whoever was guilty of a charitable action, were it only the gift of a piece of bread to a starving beggar, was to die.

Rappoport paints the Sodomites in harsh and unremitting tones. The reason for this might be explained by the exploitation of an economically vital and strategic resource, such as Dead Sea bitumen, which constituted a virtual monopoly jealously guarded by the authorities. Understandably, the wealthy traders and merchants would have been eager to protect the monopoly they held with the utmost zeal, and this zeal may have extended into broadcasting lies about themselves to discourage outsiders penetrating their domain. The supply of bitumen, and allied products into Egypt and elsewhere, was the lifeblood of Sodom – those who had no business there would have been given short shrift, in case they were interlopers, spies, or other undesirables. In this sense the uncharitable attitudes of the merchants of Sodom probably differed little from those of 'big business' in any century thereafter.

As with Genesis, we have no way of judging the veracity of Rappoport's description of Sodom or the Sodomites, and whether it was based on the Book of Jasher or from elsewhere. However, there is a basic similarity between the available accounts describing the destruction of Sodom and Gomorrah. Rappoport writes in his translation; 'what had been prosperous towns and laughing verdant valleys were now a heap of ruins, and the smoke of the country which went up as the smoke of a furnace was seen far and wide.' In Genesis (20:28) we read: 'And he [Abraham] looked towards Sodom and Gomorrah, and toward all the plain, and beheld, and, lo, the smoke of the country went up as the smoke of a furnace.'

The practice of sodomy has always been associated with sinfulness of the Sodomites, from which it derives its name, though there is no direct Biblical evidence that it existed. Homosexuality was rife throughout much of the ancient world, being more prevalent in some societies than others, and there is evidence that the fertility rites of the Sodomites, in their worship of Baal, involved both male and female prostitutes. The Book of Jasher (18:12-15) describes the revelry of the inhabitants of Sodom and Gomorrah in the following verses:

> And they had in their land a very extensive valley, about a half day's walk, and in it there were fountains of water and a great deal of herbage surrounding the water.
>
> And all the people of Sodom and Gomorrah went there four times in the year, with their wives and children and all belonging to them, and they rejoiced there with timbrels and dances.
>
> And in the time of rejoicing they would all rise and lay hold of their neighbours' wives, and some, the virgin daughters of their neighbours, and they enjoyed them, and each man saw his wife and daughter in the hands of his neighbour and did not say a word.
>
> And they did so from morning to night, and they afterward returned home each man to his house, and each woman to her tent; so they always did four times a year.

What follows in this book is an attempt to reconstruct the disaster that engulfed the 'cities of the plain' drawing on the evidence available, but since this occurred in the very dawn of history it is not without its own peculiar and unique challenges. Understandably, therefore, the contents of this book cannot be considered authoritative; neither can they be considered all-encompassing. However, it is hoped they will be viewed by archaeologists, historians, laymen, and others, as a contribution from a technological perspective to a better understanding of that age-old question – what happened to Sodom and Gomorrah?

Julian Champkin (2001) graphically summed up the cataclysm that befell the 'cities of the plain':

> With the earthquake, with the flames of burning methane and black smoke of boiling bitumen, another terror in the night for the sinners of Sodom and neighbouring Gomorrah. The very ground beneath their feet betrayed them. They fell and floundered as the ground failed to hold, while their homes sank before their eyes into the once solid earth. It drowned men, women and children, and all their livestock.

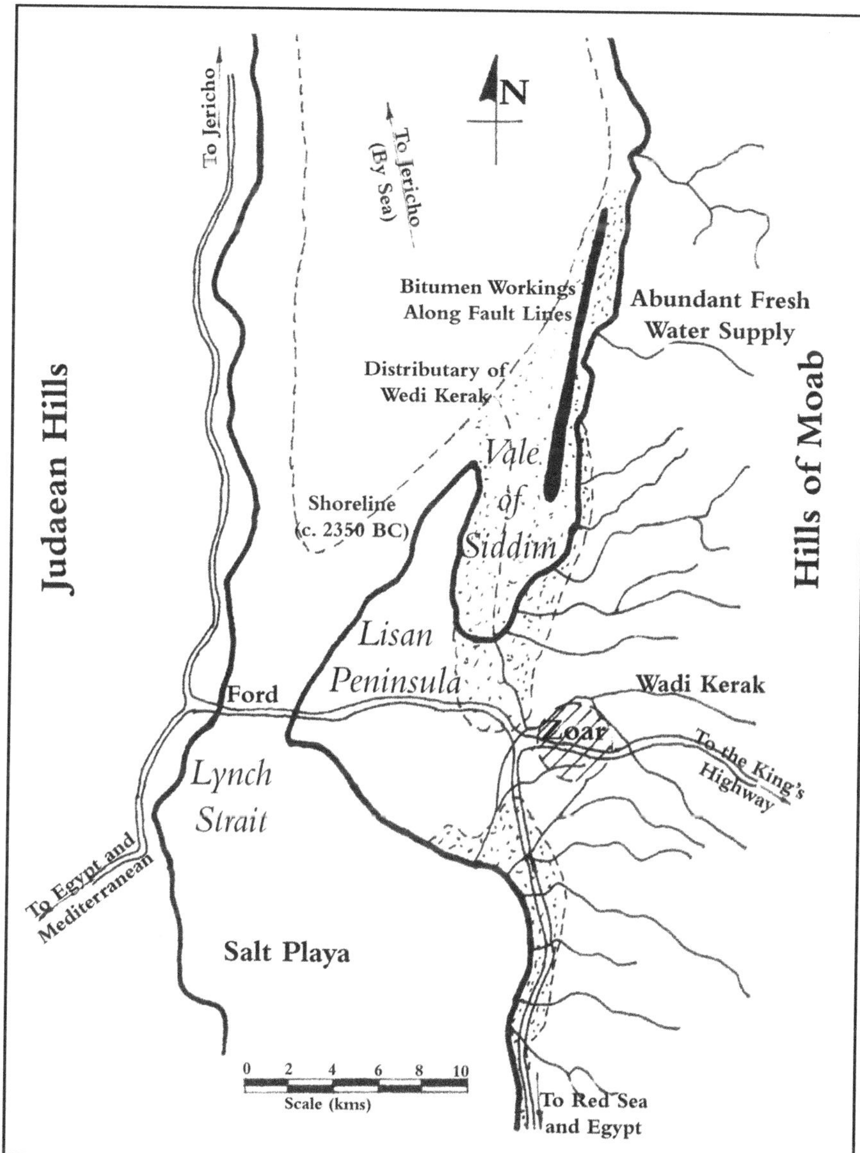

2. The postulated location of the Vale of Siddim and Biblical Zoar.

The conclusions of this book may be summarised as follows:

(1) The destruction of Sodom and Gomorrah, and the other 'cities of the plain', was an actual historical event. Reconstruction, based on scientific evidence, supports the Biblical account.
(2) The catastrophe occurred around 2350 BC, and was a consequence of earthquake induced liquefaction, the first in Judaeo-Christian

history. It involved a massive amount of material, which slid beneath the surface of the Dead Sea generating a tsunami wave which would have swept away all shoreline settlements, and struck well inland to both the north and the south. As Strabo wrote so succinctly in the excerpt reproduced earlier, 'the lake burst its bounds'. The sequence of events, from start to finish, would have lasted no more than about twenty minutes.

(3) Sodom was the centre of an industrial metropolis engaged primarily on the harvesting and processing of bitumen from the Dead Sea.

(4) The bitumen trade of the Sodomites was mostly with Egypt.

(5) The destruction of Sodom would have totally disrupted the bitumen trade, led to turmoil in commerce with its inevitable impact on regional politics, and initiated large-scale emigration out of the Dead Sea region to other parts of Canaan and beyond.

(6) The legendary Vale of Siddim, in which Sodom was sited, was located at the southeast extremity of the North Basin of the Dead Sea within the Bay of Mazra'a (Jordan).

(7) The bathymetry of the North Basin indicates the existence of a 'scar surface' in the Bay of Mazra'a, suggestive of a major underwater landslide involving a volume of about 3.5 cubic kilometres, i.e. 3.5 billion cubic metres. Debris mounds are evident at, or close to the foot of the slide in the general vicinity of Longitude 35° 31'E and Latitude 31° 26'N. This area, it is postulated, is where the remains of Sodom are most likely to be found. Though lying within Jordanian territorial waters at depths of between 300 and 700 metres, the target area is adjacent to the International Boundary, the slide debris extending into Israeli waters.

1. The Genesis Account

The Hebrew Bible

An ambitious project, termed 'The Bible Project', has been underway in Jerusalem since 1958, and is unlikely to be completed until about the year 2200. A group of scholars is sifting through all the sacred texts forming the Hebrew Bible, with the view to correcting and rewriting where necessary in order to produce an authoritative version. Matti Friedman (2011) provided a fascinating report on progress to date in his newspaper article titled 'Decades of work uncovers changes in Old Testament'. He describes how, in 1958, a group of dedicated scholars embarked on a project which none will live long enough to see to its completion; one member of the original team died in 2010 at the age of ninety years. The project has the aim of producing an authoritative edition of the Old Testament by scrutinising every alteration of text that has occurred over the centuries, and delving into every relevant extant source of religious writing.

Many Jews and Christians hold strong convictions regarding the immutability of Scripture, but the work of the scholars already has shown that many books and passages of the Old Testament have altered through the ages, evolving into the versions generally accepted today. Wearisome months and years have been spent combing through ancient manuscripts, such as the Dead Sea Scrolls, parchment texts from Egypt, and religious writings in Hebrew, Samaritan, Aramaic, Greek and Latin, amongst others. It is expected that the project will continue for the next two hundred years. Thus far, after five decades of research, authoritative versions of three of the Hebrew Bible's twenty-four books have been published. (Christians count the same books differently, for a total of thirty-nine).

It is clear from the progress to date that the texts of the Old Testament are far from inviolable, having experienced numerous changes throughout their long history and been subject to scribal errors, inaccuracies, additions and deletions, and other textual changes that, understandably, have occurred over the generations in oral communication or by the written word.

It remains to be seen whether the work of the Bible Project team will shed further light on the Genesis account of the destruction of Sodom and Gomorrah. Perhaps more substantial evidence will emerge to aid in the reconstruction of those graphic events that occurred so long ago. In the meantime, we are in the fortunate position that preceding generations of scholars have provided us with the Bible we know and cherish today. Our advanced scientific knowledge is a tool to further future study and, as it relates to the destruction of Sodom and the 'cities of the plain', will be seen to be of powerful application.

The Book of Jasher

The origin of the Book of Jasher, also known as the 'Book of the Just Man', is obscure though reference is made to it in both the Hebrew Bible and English translations, e.g. Joshua (10:13) and Samuel II (1:18). These inclusions prove it was extant at the time the texts of Joshua and Samuel II were written. As no copy in the original Hebrew is known to exist, it may indeed represent a 'lost book' as some claim. It is to be hoped the Bible Project, currently in progress, will illuminate this obscurity.

Scholars have attempted to reconstruct the book, and several versions have been published, however caution must be exercised regarding the reliability of its factual content, the authenticity of which may be fragmentary. It is non-canonical, its substance not being considered scripturally genuine by religious authority. Despite this it is an interesting and valuable reference. Its ninety-one chapters are a condensed history of the first books of the Old Testament, extending from the Creation up to the occupation of Canaan following the death of Joshua.

The Book of Jasher makes reference to Sodom and its inhabitants in a similar manner to that of orthodox English translations, but with a notable exception discussed later in this chapter. Other verses describing the debauchery of the Sodomites in their quarterly revels have been quoted earlier.

English Versions

The first English version of the Bible can be attributed to John Wycliffe (c. 1320-84), who translated the Latin *Vulgate* with the assistance of friends. Wycliffe was proclaimed a heretic for challenging church doctrines, his works condemned, and his followers imprisoned. He himself was unmolested and permitted to retire to his rectory at Lutterworth, Leicestershire. It is fortunate for posterity that Wycliffe's translation was not burned, but remained in manuscript form until 1850 when it finally appeared in print.

The first attempt to produce a printed version in English was that of William Tyndale. The first edition of his New Testament, printed in Cologne in 1525, was denounced by the Bishop of London, and copies burned. This did not deter Tyndale from producing his translation of the *Pentateuch*, which appeared five years later. Notable other early versions include that by Miles Coverdale (1535), generally considered inferior to Tyndale's, and that by John Rogers (1539) who completed Tyndale's translation. The first authorised version was in 1540, and is generally known as Cranmer's Bible from the influence of Archbishop Thomas

Cranmer, who played an important role in Reformation politics in England. With various revisions Cranmer's Bible remained the official version until 1598, it being decreed by royal proclamation that copies be placed in all parish churches in England.

Perhaps the greatest scholarly translation of the Bible into English is that represented by the King James Bible, an authorised version which appeared in 1611. The work was undertaken by forty-seven scholars. By the accuracy of translation, and the purity of its style, it superseded all previous versions, and remains popular today through the sheer beauty of its language. A number of other versions have appeared in recent years purporting to be more accurate. However, for the purpose of this book, it is the passages in the King James Bible which will be used to discuss the events relating to the destruction of Sodom, Gomorrah, and the other 'cities of the plain'.

The Book of Genesis is the first book of the *Pentateuch*, i.e. the first five books of the Old Testament. The word 'Genesis' is derived from the first word of the text *bê rē'šît*, Hebrew for 'in the beginning'. In its initial chapters (1-11) Genesis relates the stories of the Creation, the Garden of Eden, the creation of Adam and Eve, the fratricide of Cain, Noah's Ark, the Flood, and the Tower of Babel. From this point Genesis tells the Patriarchal story of Abraham and his wanderings, his sons and family. Included in Abraham's entourage is Lot, Abraham's nephew, who was destined to survive the destruction of Sodom. The story of Lot, and his struggles to maintain a virtuous existence amidst the debauchery of Sodom, is a focal part of Genesis, and delivers a powerful moral message as Sodom was doomed, through the iniquities of its inhabitants, to be destroyed by fire and brimstone according to the Genesis account.

Kuhrt (1995) gives details of the variations in style that may be detected in the *Pentateuch*. These include narrative, priestly and authoritative styles, suggesting that the original compositions were compiled by a number of different hands. As Kuhrt points out there are numerous anachronisms and inconsistencies. These resulted in disputes in the past between theologians, priests, and scholars. Such debate continues to this day, seemingly as vigorous as it ever was in the synagogues of Old Jerusalem.

The Genesis Account

In Genesis, facts concerning Sodom are interwoven within the historical account, as it is recounted in the Old Testament. It is reasonable to question whether the writers of Genesis may, or may not, have chosen versions of the event from the oral tradition in order to provide a suitable backdrop to the moral message they wished to convey. It is expected they seized upon the opportunity if such a tradition existed.

It is only in the last century that the science of seismology has developed, and advanced to the point where we now better understand the causation of earthquakes through plate tectonics and accumulated shear stress phenomena, and are capable of measuring and recording seismic events. We may not be able to predict when and where earthquakes may strike, but it cannot be disputed that we now know far more than the ancients, who viewed everything beyond their powers of comprehension and reason as an 'act of God'.

The above should not be viewed as an attack upon the veracity of the Bible for, as has been stated earlier, the geological perspective regarding the destruction of Sodom is generally supportive of the Biblical description. The eyewitness account of the destruction of Sodom, however, as recounted in Genesis, was written in the dawn of history.

On the presumption it was a real live event, it is important to divorce the emotional struggles of Lot from those clues which enable a rational hypothesis to be developed leading up to the dramatic destruction of Sodom. Whether Lot and his family lived through the cataclysm is not immaterial, but is of secondary consideration. What is vital is to identify those passages of Genesis which impart specific information relevant to the catastrophe from those that don't. By this means we can begin to analyse this distant event using the scientific techniques and technical knowledge currently available.

Chapters 12 and 13 of Genesis describe the long wanderings of Abraham and Lot, and how they go down to Egypt. On the southward journey through Canaan an altar is built on a mountain at Bethel, an altar to which Abraham and Lot return after their sojourn in Egypt. There is trouble between the herdsmen of Abraham and Lot, and they decide to go their separate ways:

> And Lot lifted up his eyes, and beheld all the plain of Jordan, that it was well watered everywhere, before the Lord destroyed Sodom and Gomorrah, even as the garden of the Lord, like the land of Egypt, as thou comest unto Zoar. Then Lot chose him all the plain of Jordan; and Lot journeyed east . . . and Lot dwelled in the cities of the plain and pitched his tent towards Sodom. (12:10-12)

Chapter 14 relates the battle of the kings in the Vale of Siddim following an invasion by confederate armies from the north. It is a chapter of some importance, being the first chapter in the Old Testament to describe a battle. As the battle must have been an event of epic proportions, and of political prominence to warrant inclusion in Genesis, its implications are discussed in detail later in this chapter.

Lot is reported as having suffered the indignity of being made captive by the invaders and taken away, presumably destined for a life of slavery, but is rescued by Abraham from an uncertain fate. Later on, God appears to Abraham, and announces his intention of destroying Sodom and Gomorrah 'because their sin is great' (18:20), but Lot is forewarned by angels who urge him to escape the impending doom with the words 'Up, get you out of this place; for the Lord will destroy this city' (19:14). Lot's flight from Sodom is described thus:

> And when the morning arose, then the angels hastened Lot, saying, Arise, take thy wife, and thy two daughters . . . lest though be consumed in the iniquity of the city. And while he lingered, the men laid hold upon his hand . . . and they brought him forth and set him without the city. . . . Escape for thy life; look not behind thee, neither stay thou in all the plain; escape to the mountain, lest thou be consumed. (19:15-17)

Lot arrives safely in Zoar and destruction of the cities ensues:

> The sun was risen upon the earth when Lot entered into Zoar. Then the Lord rained upon Sodom and upon Gomorrah brimstone and fire from the Lord out of heaven; And he overthrew those cities, and all the plain, and all the inhabitants of the cities, and that which grew upon the ground. But his [Lot's] wife looked back from behind him, and she became a pillar of salt. (19:23-26)

Abraham, from his position on the mountain at Bethel (near Hebron), is a witness to these dramatic events:

> And Abraham gat up early in the morning. . . . And he looked toward Sodom and Gomorrah, and toward all the land of the plain, and beheld, and, lo, the smoke of the country went up as the smoke of a furnace. (19:27-28)

A detailed study of the Hebrew phraseology describing these vivid events reinforces the English translation, but there are subtle differences:

(1) Bentor (1989) points out that the original Hebrew word translated as 'smoke' is *kitor*, which signifies water vapour. This might suggest, therefore, that the primary catastrophe to which Sodom was presumably subjected involving fire and brimstone, was followed by a secondary one involving water.
(2) Wood (1999) writes: 'The Biblical description, then, of the destruction was of burning material raining down from above, accompanied by an overturning of the cities and thick smoke being forced upward from the land. A rather apocalyptic scene, one that was forever etched in the minds of the ancient Israelites.'

Wood also attempts to determine a precise date for the destruction of Sodom from the chronology of the lives of the Patriarchs – Abraham, Isaac, Jacob and Joseph. From this he arrives at a date of about 2070 BC. However, to give Wood credit, he admits to an inconsistency of 230-280 years, thereby placing it within the period 2350-2300 BC. This is a time generally accepted by archaeologists, but not without reservations, as towards the end of Early Bronze III.

A Biblical basis for the dating of the destruction of Sodom is heavily reliant upon the assumption that Abraham actually lived and witnessed the event. Naturally, we have no other source of verification. Thus there is a dichotomy. Do we take the Biblical account as factually correct and ignore science? Or, do we place emphasis on interpreting scientific and historic evidence in order to produce a rational and logical argument, which may or may not be at variance with the genealogy of the Patriarchs implied by Genesis? This book takes the second path.

It may be argued that the writers of Genesis were unable to accurately portray the early history of Israel in the period of the Patriarchs, preferring to relate a number of disparate incidents useful in imparting a moral message. Therefore, the stories as told in Genesis, and indeed elsewhere in the early parts of the Old Testament may be neither historically accurate nor factually correct. The *Pentateuch*, most assuredly, was written to preserve Israel's rich cultural heritage at the time of compilation, with emphasis upon Judaic theology, genealogy, and fixing its prescribed laws and customs pertaining to religious observance. More cynically, Genesis might be construed as offering moral justification for Israelite occupation of the Land of Canaan following the Exodus.

Tubb (1998) writes:

> It is clear that the time of writing, editing, or compiling of, say, the Patriarchal stories is of critical importance, for it obviously has a bearing on the credibility of the source material. In this respect, a number of points can be made. First, the narratives represent a compilation of very many individual sources. Second, the process of editing these sources into the coherent whole we recognize as the Book of Genesis could not have taken place before the seventh or possibly sixth century BC. Third, the writers or editors had access to a variety of material, most of which can be assumed was contemporary or near-contemporary with the time of writing.

The implications are obvious. Firstly, because of the persistence of the saga from antiquity, it is highly probable the destruction of Sodom and Gomorrah occurred a long time before it was recorded in Genesis, with the likelihood the circumstances involved had become nothing more than fragments of folklore. Secondly, the account is markedly deficient with regard to specific details of importance to the forensic investigator. Regrettably, we have no more than a few scant Biblical passages upon which to commence our reconstruction of the events described, and to put them into modern perspective using current scientific and technical knowledge.

It is significant to the argument presented herein, that following the destruction of Sodom and Gomorrah, as recounted in Genesis, there is no further mention in the Bible of the Vale of Siddim. It follows, therefore, that the Vale of Siddim was lost as a consequence of the destruction and, furthermore, disappeared without trace.

From the Genesis account the following facts may be gleaned:

(1) The Vale of Siddim could be sighted from Bethel.
(2) The valley of the Jordan was well-watered.
(3) Zoar was at a higher elevation than Sodom.
(4) The cities were destroyed overnight by agencies that included 'fire and brimstone'.

The Battle of the Kings in the Vale of Siddim

A full account is rendered in Genesis 14 of a most unusual event for such an early chapter of the Bible, namely an invasion by four confederate kings from the north, and their assault on the 'cities of the plain'. The assault and the ensuing battle in the Vale of Siddim is detailed; the combatants on each side are named, and the route taken by the invading armies is described. The inclusion of such a military confrontation, suggests Sodom ranked as being highly important in the regional politics of the age. The clash must have had consequences of epic proportions:

> And it came to pass in the days of Amraphel king of Shinar, Arioch king of Ellasar, Chedorlaomer king of Elam, and Tidal king of nations; That these made war with Bera king of Sodom, and with Birsha king of Admah, and Shemeber king of Zeboiim, and the king of Bela, which is Zoar. All these were joined together in the vale of Siddim, which is the salt sea. Twelve years they served Chedorlaomer, and in the thirteenth year they rebelled. (14:1-4)

> And in the fourteenth year came Chedorlaomer, and the kings that were with him, and smote the Rephaims in Ashteroth Karnaim, and the Zuzins in Ham, and the Emims in Shaveh Kiriathaim, and the Horites in their mount Seir, unto El-paran, which is by the wilderness. And they returned, and came to En-mishpat, which is Kadesh, and smote all the country of the Amalekites, and also the Amorites, that dwelt in Hazezon-tamar. (14:5-7)

> And there went out the king of Sodom, and the king of Gomorrah, and the king of Admah, and the king of Zeboiim, and the king of Bela (the same is Zoar) and they joined battle with them in the vale of Siddim; With Chedorlaomer the king of Elam, and with Tidal king of nations, and Amraphel king of Shinar, and Arioch king of Ellasar; four kings with five. And the vale was full of *slimepits*; and the kings of Sodom and Gomorrah fled, and fell there; and they that remained fled to the mountain. And they took all the goods of Sodom and Gomorrah, and all their victuals, and went their way. (14:8-11)[1]

---

1. Note: According to Bentor (1989) the term here translated as 'slime-pits' occurs in other translations as 'lime-pits' or 'clay-pits'. The original phrase is be'eroth hemar, meaning asphalt wells, the translators apparently having confused hemar (asphalt) with homer (clay). As will be seen in a later chapter the region is rich in asphalt (or bitumen).

The Book of Jasher (16:2) gives the invaders a mighty army, so huge that some doubt must be expressed regarding the actual numbers:

> And these four kings went up with all their camps, about eight hundred thousand men, and they went out as they were, and smote every man they found in their way.

Even by Biblical standards eight hundred thousand warriors seems preposterous. Assuming the figure to be correct and Chedorlaomer's men marched four abreast, the column of men would have stretched over two hundred kilometres. The prospect of feeding such an army on the move would have daunted the most able quartermaster! To put it in further perspective, the Allied invasion of France, June 6, 1944, placed 160,000 men on the beaches of Normandy the first day.

Budge (1884) comments upon Chedorlaomer's punitive raid in the following words:

> Some writers think that Khammurabi is to be identified with Amphrael, King of Shinar, who, with Arioch of Ellasar, Chedorlaomer, King of Elam, and Tidal, King of Goiim, invaded Palestine (Gen xiv). 'Amphrael' may well be a garbled form of the name Khammurabi, and Tidal seems to be a good Hittite name, and the mention of a king of Babylon with a Hittite ring need not surprise us. There is an Elamite ring in the sound 'Chedorlaomer', yet though Ellasar may represent Larsa, its king Arioch is difficult to identify. But it is probable that the statement in Genesis has an historical foundation.

Father Tadros Yacoub Malaty, a Coptic priest has written extensively on Genesis. In his *Interpretation of the Book of Genesis*, he states the following regarding Chedorlaomer:

> Chedorlaomer is a name that means 'Servant of Laomer', one of the gods of Elam. He was known for his violence and domineering, having swept all kingdoms of the South, submitted all the valley of the Jordan, and had control of the main road between Egypt and Damascus. After twelve years of humiliation, five kings [of Sodom] rebelled against him, and refrained from paying him taxes.

Malaty states (without supporting evidence) that Chedorlaomer was the ruler of the Persian Empire of Elam, located on the northwest coast of the Persian Gulf extending north to the Zagros Mountains, Amphrael was the ruler of Shinar, located amidst the southern regions of Babylonia, Arioch was the ruler of Assur which ranged to the north, and Tidal was the leader of the Hittites to the west. It would appear, therefore, that the invaders represented a far-flung confederacy involving most of modern Syria and Iraq, and reaching far into Turkey and Iran.

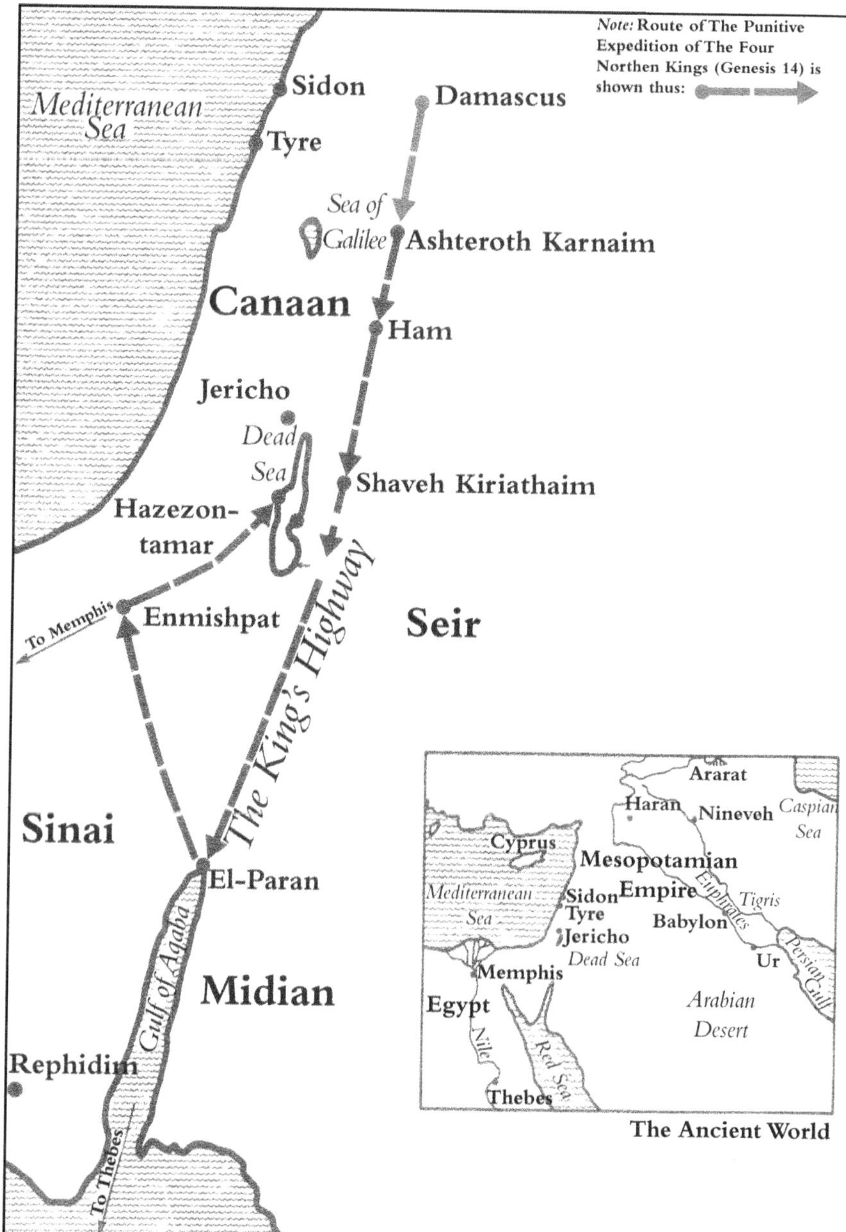

3. Route of the punitive expedition of the four Northern Kings (Genesis 14).

Figure 3 indicates the invasion route, generally accepted by Biblical scholars, as that followed by the invaders in their punitive expedition, prior to descending into the Dead Sea and engaging in battle in the Vale of Siddim. It has all the hallmarks of a military strategy designed to curtail, or control, Sodomite trade with the outside world, especially trade with Egypt. The phrase (14:4) 'Twelve years they served Chedorlaomer, and in the thirteenth year they rebelled', suggests the Canaanites were subjects who had failed to pay tribute to their overlords, and merited appropriate punishment.

The invading armies, on their march south, first took Ashteroth-Karnaim, a city in Bashan east of the Jordan River. They then captured Ham, located to the northeast of the Dead Sea, before continuing to follow what is now known as the King's Highway. This was a major route following the high ground linking Damascus with El-Paran, a town and port at the head of the Gulf of Aqaba. They then vanquished the inhabitants of Shaveh-Kiriathaim, now Kureiyat, north of Dibon in the land of Moab, before proceeding to Mount Seir, and thence to El-Paran.

Chedorlaomer's armies would have marched past the turn-off to Sodom, which would have wound westwards down through the mountains into the Vale of Siddim, and the Dead Sea beyond. If they had wanted direct engagement with the Sodomites they would have followed this route. Instead they elected to postpone the opportunity of springing a surprise attack on their enemies, which suggests they had more important issues in mind.

El-Paran was a seaport through which trade passed *en route* to Upper Egypt and beyond. It was, and continues to be, a place of considerable importance. Much commerce, in the form of imports and exports, pass through the present-day ports of Elat and Aqaba. In the days of Chedorlaomer the port of El-Paran could be expected to handle all the commerce originating in Canaan, which was bound for Upper Egypt and the Red Sea. Much of this export trade would have been in the form of bitumen and allied materials produced by the Sodomites. It might be expected, therefore, that Chedorlaomer left his tax-gatherers behind to ensure all duties and levies were paid.

After El-Paran the armies marched across the arid wastes of the Negev to Enmishpat, an oasis located approximately a hundred kilometres south of Gaza. This was situated on the overland route from Canaan to Lower Egypt. Again, Chedorlaomer is likely to have left a coterie of tax-gatherers behind to ensure his revenue collection was not compromised. He then struck off across the hills of Judaea in order to descend into the Dead Sea to Hazazon-tamar, believed to be En Gedi, on the western shore of the Dead Sea.

Chronicles II (20:2) references the conflict between the Moabites and the Ammonites with Jehoshaphat, King of Judah (873-849 BC) 'and, behold, they be in Hazazon-tamar, which is En-gedi'. En Gedi lay on the trading route winding northwards towards Jericho. It constituted a major staging-post for Sodomite trade. Once again Chedorlaomer would not have neglected to impose his control on this outpost.

At this point of the campaign Chedorlaomer and his confederates could have felt some degree of satisfaction in bottling up the trade routes, and ensuring tax revenues from Sodomite exports were fully restored. There remained only one last lesson to teach the Sodomites and their associates, and that was to thrash them in battle. We do not know what route the invading armies followed between Hazazon-tamar and the Vale of Siddim. Since the South Basin of the

Dead Sea was likely a salt plain at the time there was no need for them to march across it, or around it by following the southern shoreline. Instead they would have forded the Lynch Strait, and crossed the Lisan Peninsula, in order to besiege Sodom and her allies within the Vale of Siddim. As there is likely to have been a number of settlements along the southern shore of the South Basin, one or more of which may have been numbered among the 'cities of the plain', it is presumed that the conquest of these by the invading armies would have featured in the Genesis account if the invaders had followed this longer, more circuitous, route prior to engaging in battle.

It is impossible for the presence of Chedorlaomer's armies to have gone undetected by the Sodomites. Even though Sodom itself was largely out of sight, being within the Vale of Siddim itself, they would have had observation posts scattered throughout the region. However, the intentions of Chedorlaomer would have been obvious, and it must be expected, therefore, that the Sodomites, and their allies, would have had ample time to prepare for the inevitable battle, whether they possessed much military prowess or not.

Chedorlaomer is reputed to have been killed while his forces marched north following their successful campaign, in which they garnered much booty. His death is said to have occurred at Hoba, northeast of Damascus, and the loss of his leadership by the northern confederacies led to the Canaanites regaining their independence. It is unknown when the 'Battle of the Kings' took place. However, it must have been prior to 2350 BC, the postulated date of the destruction of Sodom.

Another important point, which may be made in connection with the Genesis account, regards the five Sodomite kings listed, i.e. the kings of Sodom, Gomorrah, Admah, Zeboiim, and Bela (which is Zoar). Strabo wrote that 'there were once thirteen inhabited cities in that region of which Sodom was the metropolis.' Figure 4 indicates twelve locations around the southern end of the Dead Sea capable of sustaining viable agriculture. It follows, therefore, that though only five kings reportedly engaged in battle against the invaders, the forces of the Sodomites were likely drawn from more than five communities. Which of those locations depicted on Figure 4 represent Sodom's confederates in the battle is, of course, open to speculation, and further study.

The Vale of Siddim and the Valley of Salt

Following the destruction of Sodom there is no further mention in the Bible of the Vale of Siddim. Its loss must, therefore, be considered total. However, the Valley of Salt persisted, and is referenced in Samuel II and Kings II in the following verses:

> And David gat himself a name when he returned from smiting the Syrians in the valley of salt, being eighteen thousand men. And he put garrisons in Edom; throughout all Edom put he garrisons, and all they of Edom became David's servants. (Samuel II, 8:13-14)

> He [Amaziah] slew of Edom in the valley of salt ten thousand, and took Selah by war, and called the name of it Jokteel unto this day. (Kings II, 14:7)

The alluvial plain of the Wadi Kerak, located at the south-eastern corner of the North Basin would have been clearly visible from Bethel prior to its loss. This would have represented one of the most expansive, and fertile, areas to the south. Though the southern extremity of the South Basin would have offered some oases of fertility clustered around the ephemeral wadis draining into the sea, these would likely have been only partially visible from Bethel.

The South Basin of the Dead Sea was occupied by salt flats inimical to large-scale human habitation and, therefore, offering a dismal vista, and beyond lay the desolate wastes of the Aravah valley. As the alluvial plain of Wadi Kerak would have been well-watered it is considered to be the prime location of the legendary Vale of Siddim, especially as the North Basin provided a ready source of bitumen.

Though ancillary to the Genesis account of the destruction of Sodom, the Prophet Ezekiel supplies comment to support the view that the singularity of the event had been long remembered. He writes (16:49-51):

> As I live, saith the Lord God, Sodom thy sister hath not done, she nor her daughters, as thou [Samaria] hast done, thou and thy daughters.
>
> Behold, this was the iniquity of thy sister Sodom, pride, fulness of bread, and abundance of idleness was in her and her daughters, neither did she strengthen the hand of the poor and needy.
>
> And they were haughty, and committed abomination before me: therefore I took them away as I saw good.

Ezekiel was born around 622 BC, and describes his calling to become a prophet when he was about thirty years of age. He prophesied the sacking of Jerusalem, which occurred in 587 BC when the Jews were carried off into captivity by the Babylonians. The presumably oral tradition of the destruction of Sodom had, therefore, lasted some fifteen hundred years before Ezekiel wrote his words. What is most interesting is the phrase 'this was the iniquity of thy sister Sodom, pride, fulness of bread, and abundance of idleness'.

The central thesis of this book is that the production of bitumen was vital to the Canaanite economy of the Dead Sea. Because the Sodomites monopolised its production, the profits of trade would have been lucrative. Their kings, princes and merchants would have waxed rich with 'fulness of bread' and, in consequence, spent their time in an 'abundance of idleness', with its perverse pleasures and distractions. Their haughtiness and disdain towards the poor and needy is little different from that reflected today by certain regimes holding a similar monopoly.

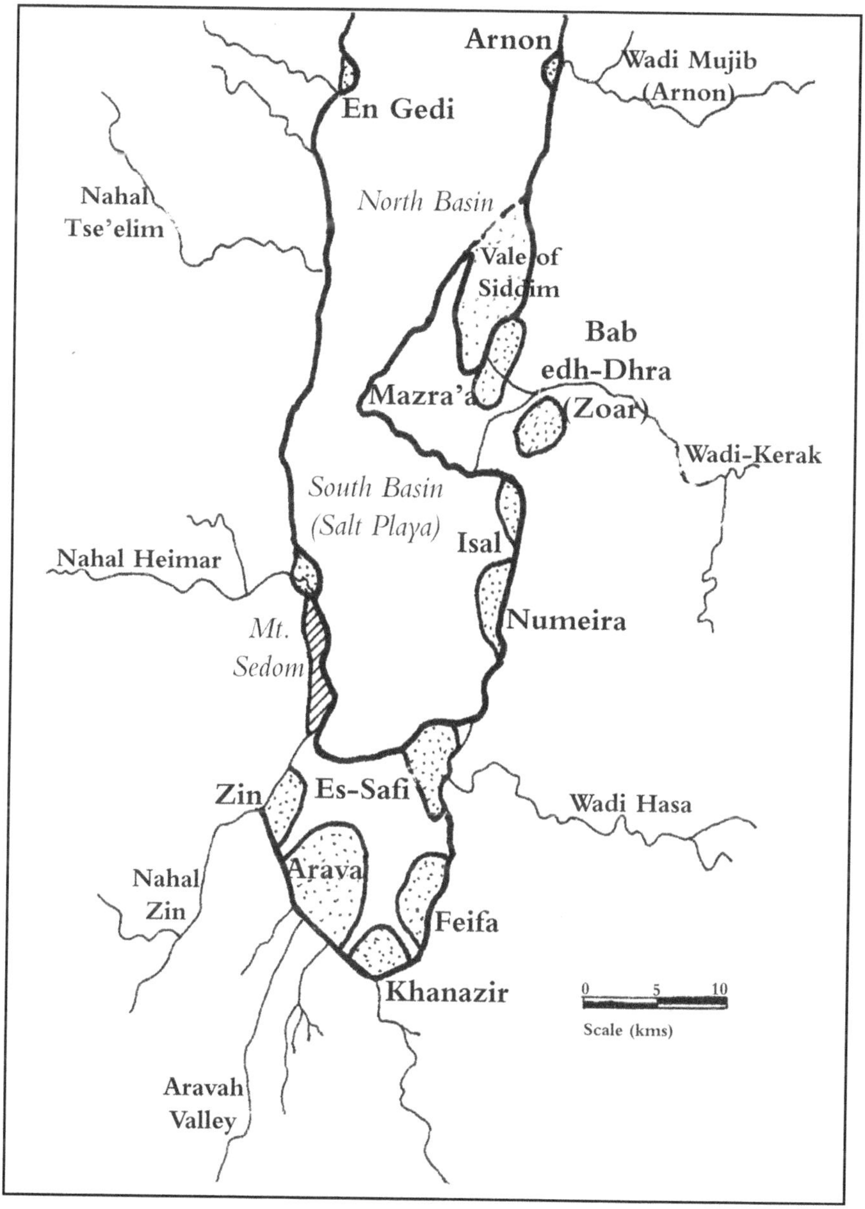

4. Southern part of the Dead Sea showing locations of agricultural communities and possible 'Cities of the Plain' (after Neev and Emery, 1995).

2. The Dead Sea

The First Survey

We now know the Dead Sea to be the lowest body of water on the Earth's surface, but this fact was not realised until 1848 after the United States dispatched an expedition to the region. Before that a good deal of myth had surrounded some of the strange happenings that had occurred there.

During the siege of Jerusalem in AD 70 the Roman commander, Titus, sentenced a number of slaves to death and had them bound in iron chains and cast into the lake. But the men did not drown, they drifted back to shore. No matter how many times the Romans tried to drown them the same thing happened. This inexplicable event made a vivid impression on the Romans, who spared the men's lives, likely crediting their salvation to the intervention of a local god.

Flavius Josephus, the Jewish historian, repeatedly referred to the Dead Sea as *Lacus Asphaltitis* and, like Strabo, stressed the fact that poisonous gases rose from its surface in all parts. Arabs added to this unsavoury reputation by claiming the gases were so harmful that no bird could fly across it, and any bird who tried dropped dead into the water. The fact the lake contained no fish, or any other living creature added to its sinister reputation.

In the autumn of 1848 boats were landed at Akko, then a small coastal community on the Mediterranean. Under the supervision of Captain William Francis Lynch, leader of the United States expedition, a long overland journey to the Sea of Galilee began, the expedition's impedimenta being hauled by teams of horses. Their objective was to travel down the Jordan River from the Sea of Galilee, and then explore and survey the Dead Sea. It took three weeks to reach the Galilee, and after Lynch set up his theodolite he was amazed to discover the water level of the Sea of Galilee was below that of the Mediterranean by over 200 metres.

From Tiberias the expedition's boats followed the meanderings of the Jordan River, the initial lushness of the vegetation becoming sparser as they progressed downstream. Eventually they reached the open vastness of the lake and their first goal – Jericho. They attempted to take a refreshing dip in the waters, but to no avail. It was impossible to sink, the old stories were true. There was no sign of life either, and the air was heavy and tainted with odours of oil and sulphur.

For twenty-two days the expedition's boats criss-crossed the lake taking soundings, and Lynch remarked, 'the waves strike the sides of the boats like blows from a hammer. But the weight of the water is such that a short time after the wind has died down the sea is calm again.'

The report of the United States Expedition of 1848, compiled by Lynch (1849), is one of the great classics of exploration.[1] Though Lynch's account of 1849 contains the first scientific body of information on the bathymetry of the Dead Sea, many Europeans had visited the region in previous centuries and some had recorded their observations. During the nineteenth century the Holy Land became a very special place of interest to travellers, many of whom went on pilgrimage to the Holy sites, and then availed themselves of the opportunity to venture into remote, and occasionally dangerous, regions, the Dead Sea included. Apart from Lynch others returned from their explorations to record their observations, including Seetzen who made a complete circuit in 1806-07; Irby and Mangles (1818), Kinglake (1835), de Saulcy (1851), Tristram (1864), duc de Luynes (1864), Allen (1850) and Wilson (1865). Some unfortunates, like Christopher Costigan (1835), and Thomas Molyneux (1847) perished in their endeavours.

Geography

The surface of the Dead Sea lies at over 400 metres below sea level (about 417 metres below mean sea level in 2006). Its setting is harsh and desolate, with scant vegetation cladding its slopes apart from a few locations where the presence of fresh water has created oases close to the shore. It lies within one of the world's most notable topographic features, the Great Rift Valley, a crack in the Earth's crust which stretches from Syria southwards to Mozambique, a total distance of some five thousand kilometres.

Each year the Dead Sea is visited by large numbers of people attracted by the reputation of its salt-laden waters to heal afflictions such as psoriasis, arthritis, and other common afflictions. Its waters are so dense, possessing dissolved salt contents higher than any other large body of water that it is impossible for a human to sink. Strabo (XVI, 2.42) wrote:

> [I]t has water so very heavy that there is no use for divers, and any person [who] walks into it and proceeds no farther than up to his navel is immediately raised afloat.

The bordering escarpments consist of fault scarps or steep monoclines, marked by cliffs, the steepest being on the western side. Both border escarpments are marked by over thirty horizontal shoreline terraces, or relic gravel bars, varying between about −180 metres and −400 metres MSL.

1. Modern day bathymetric charts incorporate data collected by Lynch and his colleagues, including that of the Geological Survey of Israel (2000).

The present Dead Sea has an overall length of about 80 kilometres from the mouth of the Jordan River in the north to Es Safi in the south, being divided into two parts – the North Basin and the South Basin. The Lisan Peninsula, which juts out from the eastern shore, separates the two basins and constricts the width of the Dead Sea at its narrowest point to less than five kilometres. The North Basin is by far the deeper of the two basins with water depths up to 400 metres, whereas the South Basin had a water depth of less than ten metres prior to its development as a sequence of evaporation ponds for the chemical industries. The total volume of water contained within the Dead Sea approximates to 135 cubic kilometres.

Our current knowledge of the bathymetry of the Dead Sea is somewhat limited having been complicated by Israeli-Jordanian politics in the past, the international boundary between the two countries dividing the two basins almost equally. Much of our knowledge of the bathymetry of Jordanian waters, especially in the North Basin, relies on older soundings, particularly those made by Lynch in his survey of 1848, supported by more recent coring operations conducted by Elazari-Volcani (1943a, 1943b), whose studies related to identifying the bacteria of lake bottom sediments. On the whole, our bathymetric knowledge within Israeli waters of both basins is the more comprehensive as a result of the work by Neev and Emery (1967). This followed on the work of Emery (1962) who performed soundings in the Jordanian section of the South Basin. Bathymetric contours are, therefore, compiled from a number of sources and, obviously, would benefit from more systematic treatment.

The Lisan Peninsula is one of the most prominent features of the Dead Sea Basin. Jutting out from the eastern shore it has the shape of a tongue, and so is aptly named, for 'lisan' in Arabic means 'tongue'. Its high point has an elevation of −295m MSL. It is composed entirely of a sequence of calcareous and gypsiferous sediments with layered inclusions of salt, sands, gravels, and other associated sediments. The shoreline around the Lisan is dominated by a series of low bluffs, with a top elevation generally coincident with the −375m contour.

One of the earliest accounts of the Lisan Peninsula is that contained in the travelogue of Irby and Mangles (1823). They had journeyed to the Dead Sea from Kerak, and had encountered a caravan which they accompanied part of the way, separating from it after reaching the Bay of Mazra'a. While they explored the north shore of the Lisan Peninsula the caravan continued westwards towards the strait separating it from the western shore. They wrote:

> The western horn of the bay is formed by a sharp promontory [Cape Costigan], projecting forward into the sea, in a direction nearly from south to north; that is to say, such is the relative bearing of the extremities, for there is a concavity in the line of shore where the salt-water stagnates and evaporates [*see Plate V*]. We found several persons peeling off a [solid] surface of salt, several inches in thickness; they were collecting it and loading it on asses.

> The strait is formed by a long promontory projecting from the opposite or western shore; and fortunately, just as we arrived at the narrowest part, where the ford is indicated by boughs of trees, we observed the small caravan from Karak landed on the opposite side; and as we could discern the species of animal, as well as the people on their backs, we were all agreed in estimating the distance about a mile. As there were asses of the party, the depth cannot be great.
>
> Leaving the narrowest part of the strait, we followed it to its southern extremity [Cape Molyneux], where it opens into the back-water [South Basin], and about the shore of the back-water itself to some distance. The high water-mark was at this season [June 1818] a mile distant from the water's edge. We were told this back-water is never dry, and that the ford is not at any season impassable.

Klein (1965) writes about a stone cairn, about 75 centimetres high, that appeared to represent a marker for the ford across the strait between the western shore and the Lisan Peninsula, often referred to as the Lisan Strait but known more commonly as the Lynch Strait (Fig 5):

> A small cairn was found at the mouth of one of the main arms of the Mor delta (co-ordinates 1859/0748) opposite the path on the Lashon [Lisan] in the summer of 1958.
>
> As there were no other such cairns found on the Israeli shore of the Dead Sea, we assumed that this cairn was intended to show the way eastward to the ford for the caravans coming from the western highlands.

When Lynch surveyed the Dead Sea in 1849 he wrote of the Lisan Peninsula:

> I named its northern extremity 'Point Costigan', and its southern one 'Point Molyneux', as a tribute to the memories of two gallant Englishmen who lost their lives in attempting to explore this sea.

Christopher Costigan was an Irishman who, in 1835, was the first modern adventurer to explore the Dead Sea by boat. He perished from thirst and heat exhaustion on the point now bearing his name. Thomas Molyneux, a Royal Navy officer, met a similar fate in 1847, immediately prior to the United States expedition into the region. It was a gracious act on the part of Lynch to thus honour his immediate predecessors who had perished so tragically.

Dead Sea Water Levels

Figure 6(a) indicates the considerable variation in level of the Dead Sea over the past 35,000 years, as determined by Frostick and Reid (1989), which includes the work by Klein (1982) and Begin *et al.* (1985). However, Lake Lisan, the

5. Stone cairn on the western shore of the Lynch Strait marking the ford (Klein, 1965).

predecessor of the present Dead Sea, existed long before then, and is considered to have had a continuous presence during Pleistocene times from between 70,000 to 12,000 years BP. The highest stand of Lake Lisan was about −180m MSL, fluctuating for about ten thousand years, before dropping rapidly by about 500 metres to −700m MSL between 14,000 and 13,000 BP. Since then it rose just as rapidly. More recently it has fluctuated about −400m MSL.

Figure 6(b) indicates variations reported by Sauer (1978), between the years 1800 and 1980. It may be noted that the lake rose during the nineteenth century, attained its highest level of about −390m MSL, maintaining it for approximately thirty years, since when it has fallen steadily. Between 1960 and 2010 the level has dropped, on the average, by about a half metre each year, but this annual drop has doubled in more recent years to about one metre per year.

The current level (2011) is approximately −423 metres MSL. The main cause for this decline may be attributed to the reduction of inflow of fresh water into the Dead Sea. Due to the insatiable demand of fresh water for irrigation and water supply to meet burgeoning populations in both Jordan and Israel, flow in the Jordan River has been reduced to a trickle. Furthermore, chemical plants have extended their operations over the entire South Basin and converted it into one vast evaporation pond. Whereas in the past the salt-laden feedstock water flowed freely into the first of several sequential evaporation ponds, it now has to be pumped.

Calcium	16.86	Chloride	212.40
Magnesium	40.65	Bromide	5.12
Sodium	39.15	Sulphate	0.47
Potassium	7.26	Bicarbonate	0.22

Table A – Composition of Dead Sea water in grams per litre (after Neev and Emery, 1967)

Sauer (1978) notes that since 1929 the level of the Dead Sea has diminished in volume from 144 cubic kilometres to 137 cubic kilometres, an overall reduction of about 5 percent during the period. Although the changes in water level appear dramatic, having significant impact on development along the

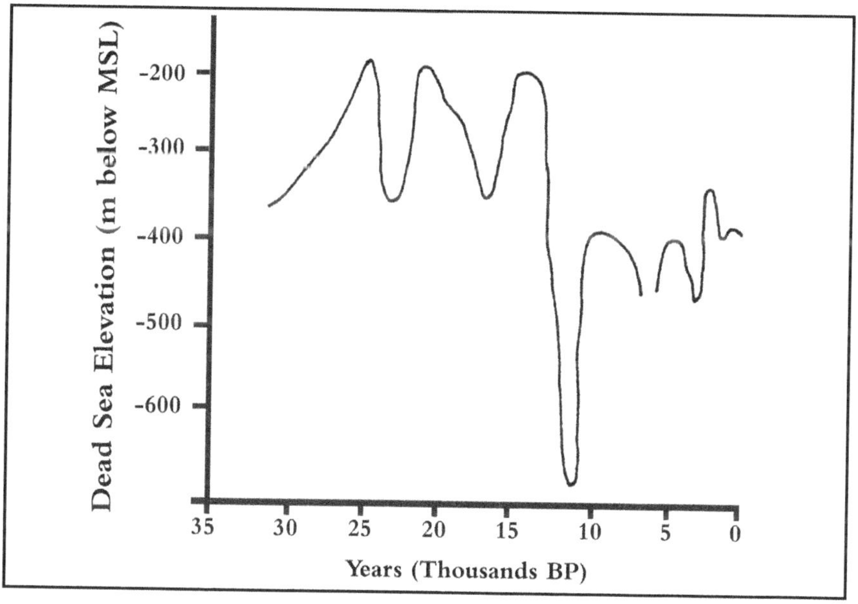

(a) Past 35,000 years

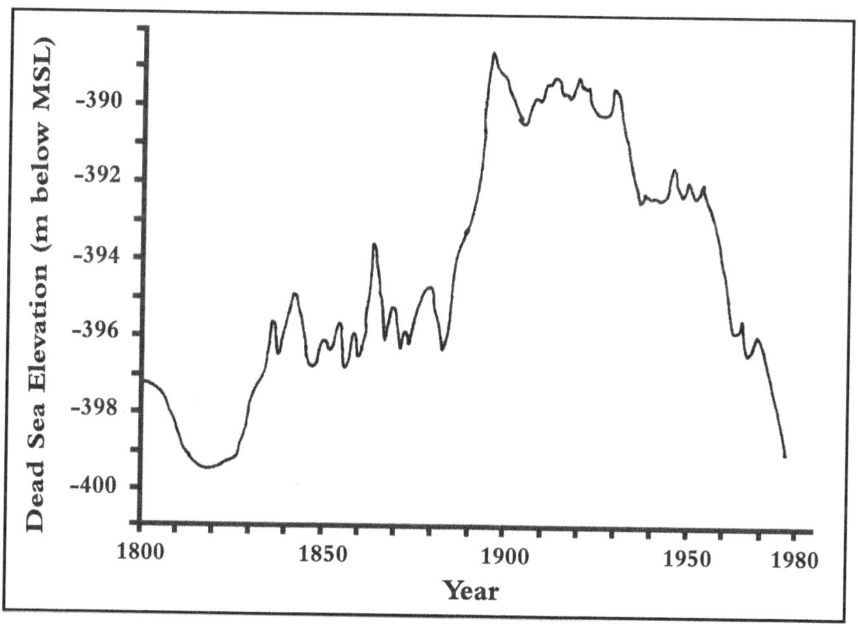

(b) 1800 to 1980

6. Variation in Dead Sea levels (a) past 35,000 years, and
(b) 1800-1980 (after Frostick and Reid, 1989, and Sauer, 1978).

shorelines, the changes observed are not beyond the bounds of expectation in an area subject to high evaporation, low variable inflow, and increasing industrial demand.

Overall Water Balance

The total catchment area of the Dead Sea is 43,000 square kilometres. The main contributors are the perennial Jordan and Yarmuk River systems in the north, and the ephemeral Nahal Arava and Nahal Zin in the south. Before water diversions took place within the Jordan valley the combined contribution of surface and groundwater inflows into the Dead Sea was of the order of 1980 Mm^3/year, of which 1670 Mm^3/year was from surface sources. Surface inflows from the Jordanian side of the Dead Sea were almost four times greater than from the Israeli side, and this is likely to have pertained during antiquity. However, since the 1960s various water diversion schemes have been initiated by both governments, with the result that surface inflows have fallen significantly to about 147 Mm^3/year from Jordan, and 13 Mm^3/year from Israel at the time of writing. The consequence has been a serious drop in the level of the Dead Sea with its volume having been depleted by 37.6 billion cubic metres. This has led to an increased lowering of groundwater levels in adjacent areas, with the likelihood that salinity will occur more rapidly in the future.

In recent years a project to stabilise Dead Sea levels has been given serious consideration by supplementing water inflows from the Mediterranean. Not only would this restore the imbalance caused by the draw-off of water for irrigation, but would enable useful quantities of hydro-electric power to be generated. Studies by Sauer (1978) indicate the scheme would have negligible effect on chemical, physical, and ecological aspects of the Dead Sea. Water-budget calculations indicate a maximum diversionary input of 700 Mm^3/year would not significantly affect long-term average water levels. The average inflow into the Dead Sea from all sources is estimated to require some 900 Mm^3/year to stabilise lake levels.

Geology

About 66 million years ago, during Upper Cretaceous times, an extended Mediterranean Sea covered the entire region, and calcareous sediments accumulated in vast quantities in a tropical to subtropical environment. These were to become the limestones which are the dominant rock type of the region. As a consequence, sea depths decreased and silica-rich muds began to be deposited, which later consolidated to form layers of chert often with thicknesses of up to thirty metres. Where the chert remains as a capping to the softer, less resistant limestone below, the topography is rugged and picturesque, with dramatic steep-sided gorges incised into the less resistant underlying limestone.

7. A typical gorge incised into limestone terrain.

The subsequent Miocene Epoch occurred between 23 and 5 million years ago, during which major crustal movements and ruptures of tectonic plates occurred resulting in mountain-building in various parts of the globe. This period of upheaval also helped develop the numerous rift valley systems we know today, including that of the Great Rift Valley, which runs from Mozambique to Syria, of which the Dead Sea forms a conspicuous feature towards its northern extremity.

The border escarpments are infested by fault scarps resulting in cliffs, steep slopes and narrow canyons incised into the predominantly Cenomanian and Turonian limestones and associated strata on the western side of the basin. The eastern slopes are composed of softer Nubian sandstones of Paleozoic to Middle Cretaceous age. The complexity of the rock strata on both sides of the Dead Sea can be attributed to graben formation, as much geological structure is evident in the form of subsidiary faulting. There are also outcrops of volcanic rocks in the form of plugs and flows widely distributed along the Eastern Escarpment Fault. These are considered to be of Quaternary age. Beyond the immediate confines of the Dead Sea there exist vast areas covered by basalt and associated volcanics. Both escarpments are mantled by over thirty horizontal shore terraces or relic gravel bars which reflect previous lake levels.

The most prominent geological feature in the region is the long fault graben, see Figure 8, whereby the Dead Sea is bounded to the east and west shores by major faults, known respectively as the Eastern and Western Border Faults. Relative movements between the bounding Arabian and Palestinian tectonic plates resulted in substantial displacement by strike-slip movement amounting to about one hundred kilometres, the Arabian plate moving northwards relative to the Palestinian Plate.

Much more is known about the Miocene from the sedimentary deposits and their included fossils elsewhere in the world, but our knowledge is deficient as far as the Dead Sea is concerned, the absence of fossil remains suggesting the region was hot and arid during the period and, consequently, hostile to life. Wind was the major erosional agent and some transportation and redeposition of sediments occurred, in conjunction with spalling of the chert rocks from major diurnal fluctuations, chert being particularly susceptible to fragmentation by this process. This paved the way for the severe erosion of the landscape that was to take place in the ensuing Pleistocene, leaving the topography much as we know it today.

At the onset of the Pleistocene, some 20,000 to 30,000 years ago, world sea levels were somewhat higher than today, but this is unlikely to have had much effect on the Dead Sea as there is a drainage divide (rising to 220 metres above MSL) in the Arava Valley to the south, being its extension towards the Gulf of Aqaba. Though the evidence suggests that during much of this period Dead Sea levels were almost 200 metres higher than they are today, the Dead Sea, in the form of its predecessor Lake Lisan, would have remained within a closed basin, the waters backing up northwards along the Jordan valley as far as the Lake of Galilee. This isolation resulted in little impact on water levels in the Dead Sea, as there was no communication with the external sea. However, the cooler climate prevailing during the Late Pleistocene was marked by a general lowering of sea levels worldwide, and this resulted in increased precipitation. Where concentrated run-off flowed down the steep precipitous flanks of the Rift Valley it incised deep gorges into the limestone rock, and eroded great quantities of material to be deposited around the shores of the Dead Sea in the form of well-sorted strata, generally granular in origin, comprising sand, gravels and boulders.

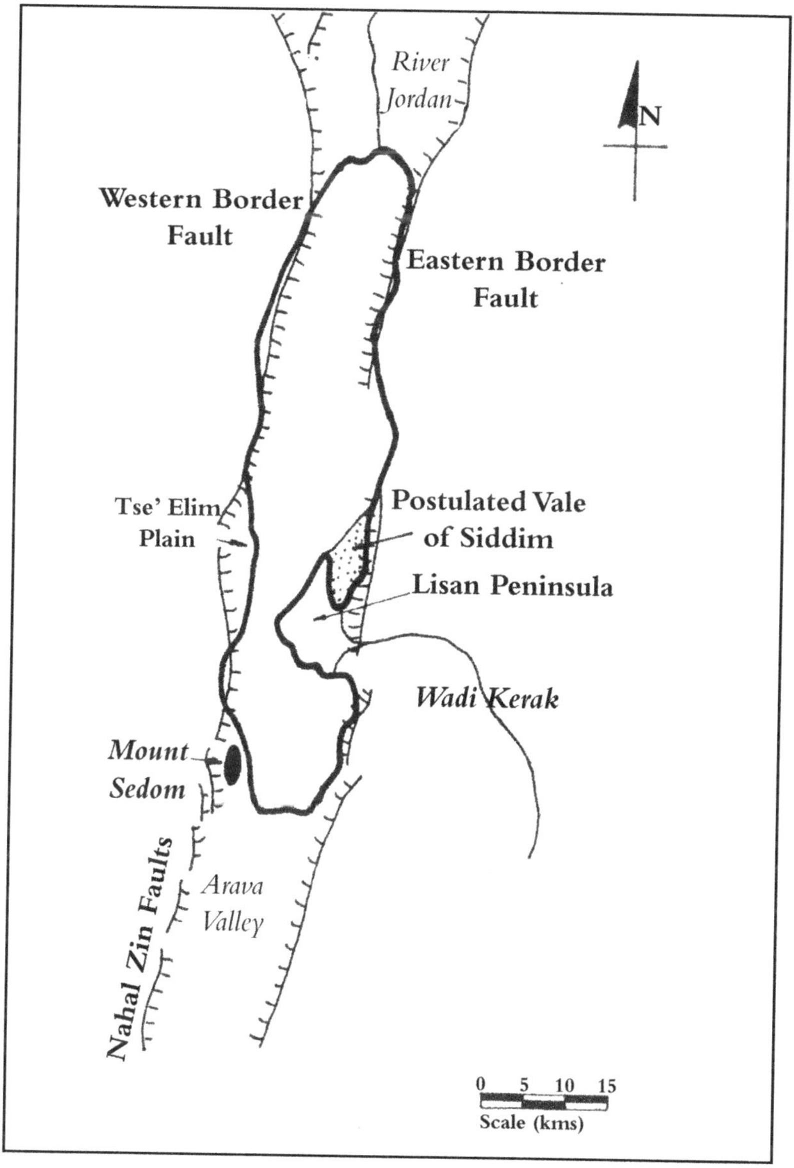

8. The Dead Sea graben and border faulting.

Some of these deltaic deposits became weakly consolidated or cemented by varying degrees of calcification, into sandstones and conglomerates of localised distribution and areal extent.

Towards the end of the Pleistocene hot arid conditions began to return to the region with the result that active erosion slowed and lake levels began to drop as evaporation exceeded replenishment. With the lowering of water levels in the Dead Sea, sequences of marl and evaporites such as salt and gypsum accumulated.

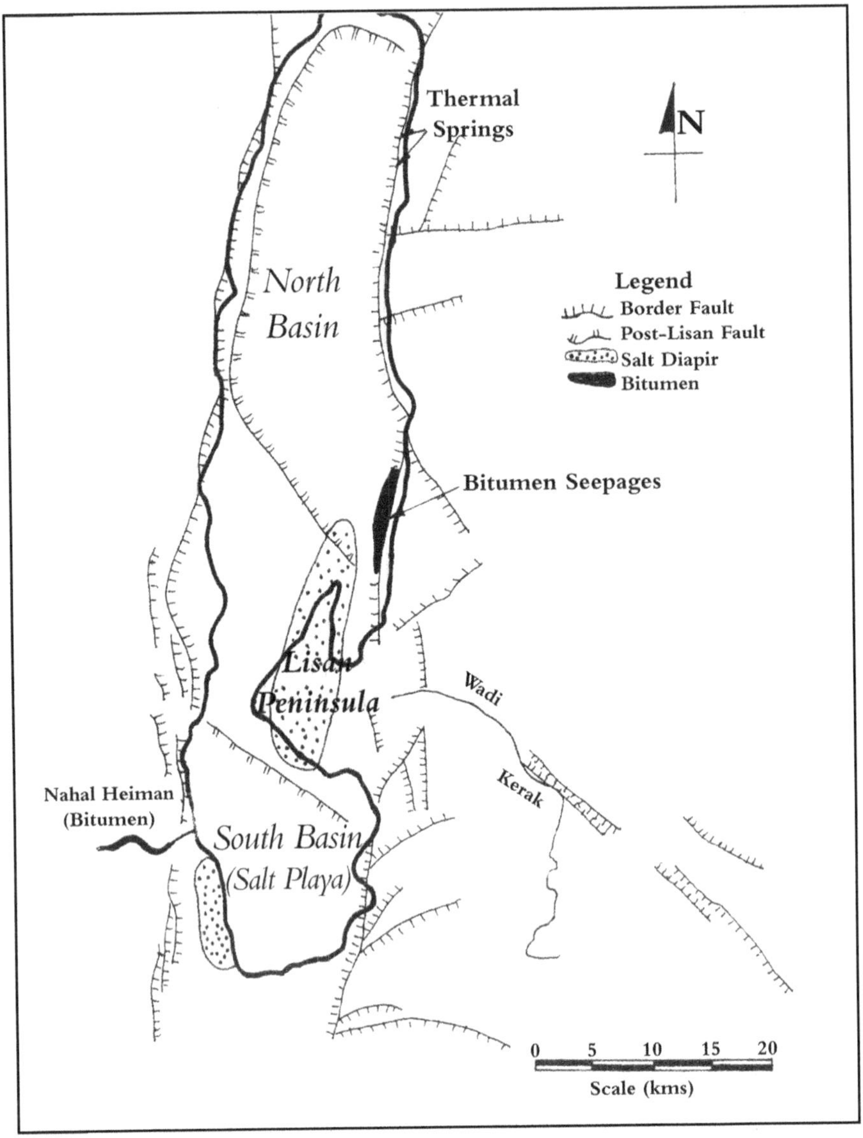

9. Salt diapirs and known bitumen locations (after Frostick and Reid, 1989).

It is estimated that beneath the Dead Sea there is a vast sequence of sediments, amounting to between three and five kilometres in overall thickness, amidst which are substantial beds of salt, the origin of which is obscure. The greatest thickness of salt proven by drilling so far is of the order of 3000 metres. This may be a diapir, as such intrusive formations occur elsewhere in the vicinity, the most noticeable being Mount Sedom on the west side of the South Basin. Figure 9 indicates the locations of the two salt diapirs in the area according to Frostick and Reid (1989). Frequently, salt diapirs are a target for oil and gas exploration, as the upthrust of the host rocks during their creation results in a disruption of strata favourable for the accumulation of migrating hydrocarbons. It may be noted that the known bitumen occurrences in the Dead Sea are in close proximity to salt diapirs.

More recent sediments comprise varying thicknesses of clays, silts, sands and gravels interbedded with varying thicknesses of crystalline rock salt. The inclusions of buried salt strata are generally thicker around the fringes of the lake, where they have accumulated as a result of evaporation. They tend to predominate in arms of the lake where there is no nearby stream bringing freshwater into the immediate area, and especially where the water depth is shallow and evaporation more intense.

Sinkholes in Exposed Beaches

The dramatic fall in the water level of the Dead Sea over the past twenty years or so has led to the development of sinkholes at a large number of locations as a result of solution of buried salt strata (see Figure 10). Some of the sinkholes that have developed are of great size and constitute a hazard in those areas of shoreline newly exposed through the falling water level. Figure 11 shows one part of a massive sinkhole that was initiated by a leaking water-pipe irrigating a banana plantation at Mazra'a, Jordan.

Arkin and Galat (2000) assert that the menace of sinkholes in exposed beaches became a major concern in 1990 when a series of holes, between two and fifteen

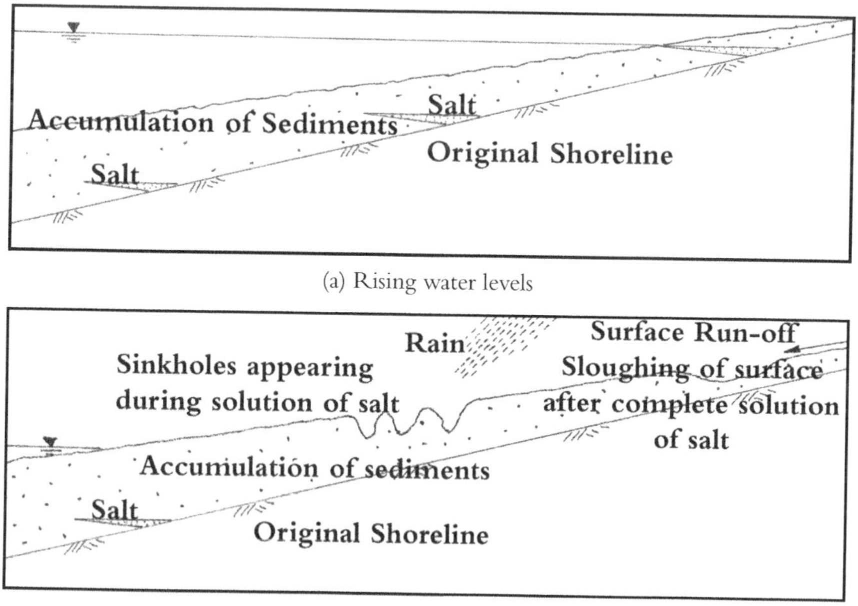

(a) Rising water levels

(b) Falling water levels

10. Effects of (a) rising water levels, and (b) falling water levels on salt deposition and solution.

metres in diameter, and up to seven metres deep, appeared in the Newe Zohar area on the western shore. One sinkhole was opened up by a bus travelling along a paved stretch of highway. Repeated infilling, further collapse, and reappearance of the sinkholes, demonstrated the extent and severity of the phenomena.

11. Edge of a massive sinkhole at Mazra'a, Jordan, resulting from leakage from an irrigation pipe.

Long stretches of the shoreline are affected, with numerous sinkholes appearing on both sides of the Dead Sea. One of the major causes can be identified with the receding water level, leading to exposure of buried salt layers along the shoreline to solution by surface runoff water. However, subsidiary causes may prevail. These could be related to internal erosion, and loss of fine particulate soils, after being subjected to increased hydraulic gradient induced by groundwater lowering. The subject of sinkhole development on both shores is complex and governed by circumstances peculiar to the locality concerned.

An interesting illustration of this extensive, and widely pervasive problem, was encountered after the commissioning of an extension of the Dead Sea Works chemical factory at S'dom in 1982. The extension was constructed on the outer fringes of an alluvial fan. Soil investigations prior to construction had failed to detect the existence of a thick layer of crystalline salt buried within the alluvium. When fresh water was used to wash down paved areas of the plant the water migrated downwards through cracks in the pavement, and dissolved the crystalline salt below causing extensive settlements amounting to a half metre or so. Only a workshop building suffered distress initially, but prompt action identified the cause and the extent to which other, more vital structures, were vulnerable. Remedial measures were quickly initiated to forestall repetition of the phenomena which might have disrupted potash production.

Closson *et al.* (2004) carried out studies into subsidence features on the Jordanian side of the Dead Sea after The Arab Potash Company had experienced problems when a section of dyke failed. The event followed the filling of a newly completed evaporation pond. They postulated the failure could be due to water pressure lubricating fractures within the soil/rock strata beneath the new pond. The adjacent Dead Sea Works evaporation ponds, on the Israeli side of the border, do not appear to have experienced a similar problem attributed to the same cause.

Dead Sea Sedimentation

As mentioned earlier, the first extensive body of water to have occupied the Jordan-Arava graben was Lake Lisan. Begin *et al.* (1974) suggested this existed between 60,000 and 18,000 years ago with its highest level reaching −180 metres MSL towards the end of this period. During this long length of time water levels fluctuated widely, the lowest level considered to have fallen to about −520 metres MSL, following which the level stabilised at about −400m MSL (Frumkin *et al.*, 1991, Neev and Emery, 1995). The nature of the sediments deposited during this period not only consisted of detritus of terrestrial origin, e.g. particulate sands and silts as a consequence of surface run-off, but lacustrine deposits due to chemical imbalance of the various salts in solution. The nature and composition of the sediments beneath the waters of the Dead Sea thus represent an important historical signature respecting past climates, and is of immense value to research scientists. Not surprisingly, calcium carbonate and calcium sulphate, in their varied forms, dominate the underwater geology.

The lowest levels attained would have virtually dried up the Dead Sea entirely. Neev and Emery (1967) gave their interpretation of climatic variation over the past 100,000 years in the form of ratios of run-off/evaporation, which is considered to mimic water levels. This has been given a broad climatological interpretation and its relation to archaeology. The results of their studies are reproduced in Figure 12, upon which has been indicated the likely period of formation of the Sedom Clay, an interesting deposit of micrite reported by Adler and Harris (1983) which, by its nature, may be considered to be widely distributed within the confines of the Dead Sea. Though the Sedom Clay is referred to later in this chapter, its properties and significance, as it relates to earthquake susceptibility is discussed in greater detail in Chapter 7.

It may be seen from Figure 12 that a prolonged dry period existed for approximately a thousand years, between around 4500 and 3500 BC, which can be identified with a cultural break between the Neolithic and Chalcolithic. A second, perhaps more significant cultural break, occurred between the Early Bronze and Middle Bronze eras. This occupied a period of about three hundred years from 2300 to 2000 BC. In both these periods it may be expected that water levels within the Dead Sea basin shrank significantly, with evaporation far exceeding supply.

Druckman *et al.* (1987) proposed a rapid fall of Lake Lisan after 14,000 BP based on ^{14}C dating of detrital plant remains in oolitic sediments south of the Dead Sea. Based on contours of submarine canyons on the seafloor of the Dead Sea, which could not have developed underwater, Neev and Emery (1967) suggested an almost complete desiccation of the lake, thus ending the Lisan phase. This is supported by massive salt beds deposited in the southern basin prior to 9850 ^{14}C years BP (Neev 1964) and in the southern part of the northern basin,

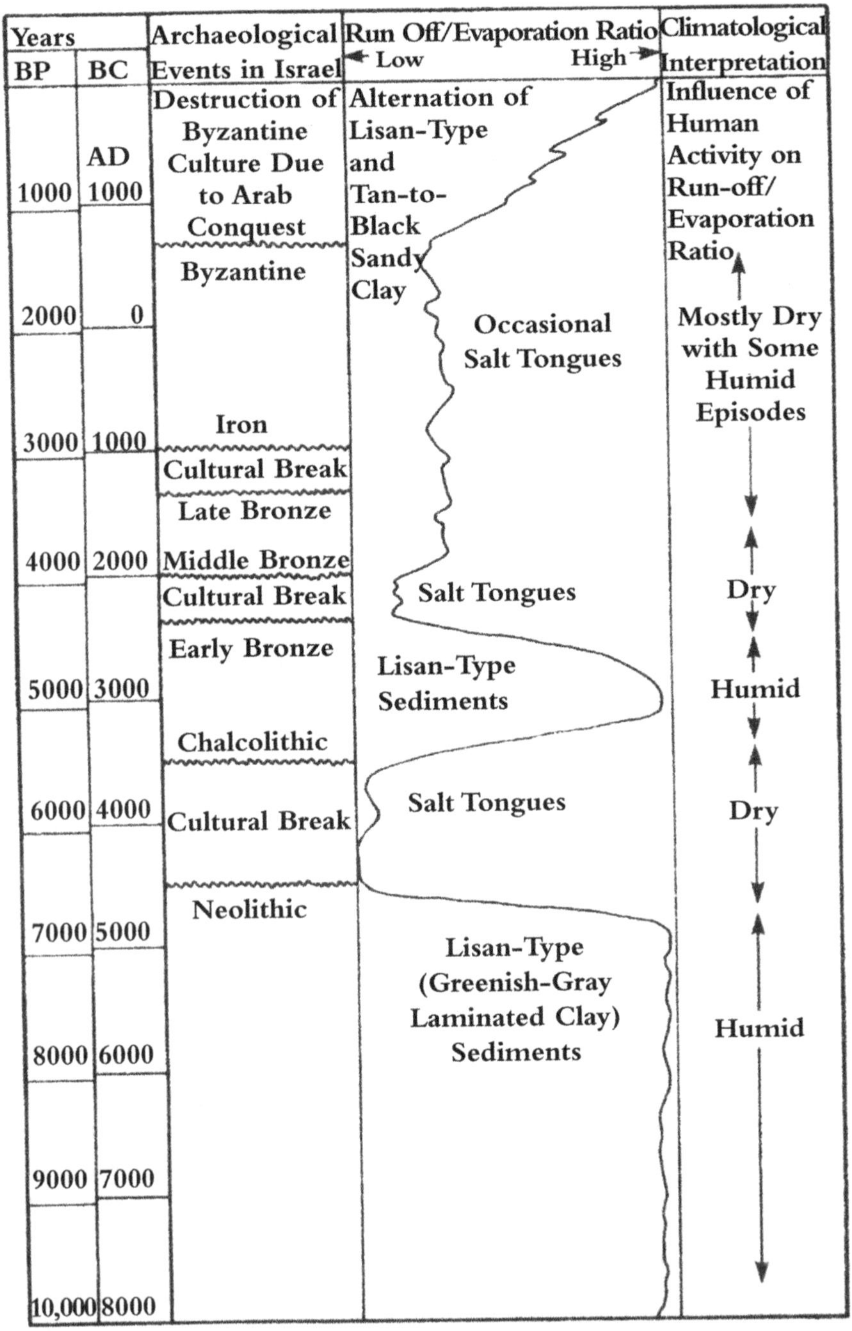

12. (a) Dead Sea stratigraphy
10,000 years BP to present
(after Neev and Emery, 1967, and Harris and Beardow, 1995).

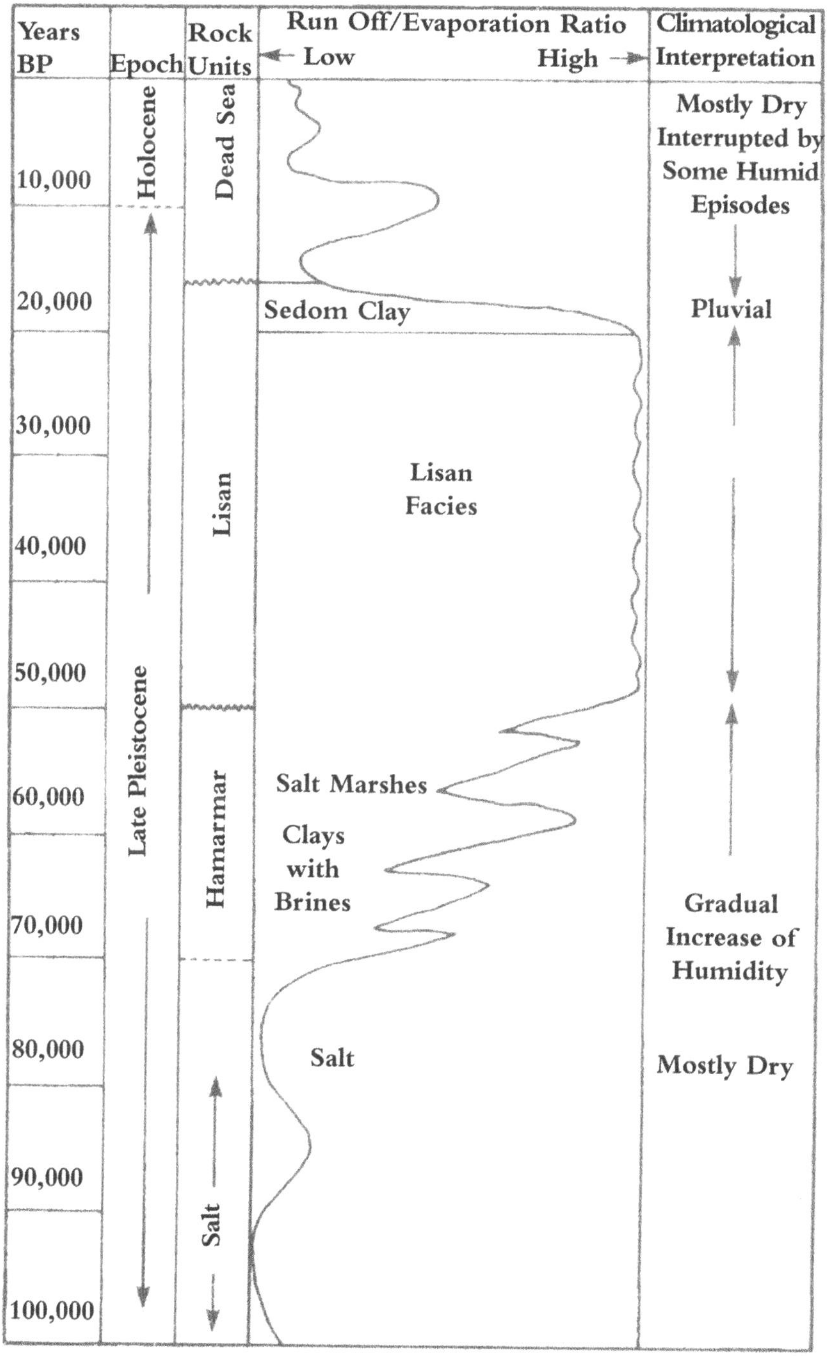

12. (b) Dead Sea stratigraphy
100,000 years BP to present
(after Neev and Emery, 1967, and Harris and Beardow, 1995).

between 11,300 and 8400 ^{14}C years BP (Yechieli *et al.*, 1993). Later Holocene sediments encountered in boreholes within the northern basin do not include massive salt beds (Stiller *et al.*, 1988; Yechieli *et al.*, 1993). Based on these dates, the extremely dry event marking the end of Lake Lisan took place between 11,000 and 10,000 ^{14}C years BP.

Frumkin (1997) comments as follows regarding the general history of water levels during the more recent past:

> Lake Lisan, the precursor of the Dead Sea, covered the Mount Sedom diapir during the Late Pleistocene and deposited the Lisan Marls on top of it (Zak, 1967; Begin *et al.*, 1980). During the rising of the diapir, the top of the salt was dissolved, creating a residual cap rock up to 50 metres thick.

Neev and Emery (1995) write:

> During the wet climatic interval of Pre-Pottery Neolithic B (PPNB), the Jordan River gorge was gradually eroding through sills that dammed both the Sea of Galilee and Beth She'an basin on their south sides. By then, 7800 BP or the beginning of Pottery Neolithic, Climatic Phase II was initiated. During the transition from Boreal to Climatic Optimum or Atlantic Interval (Gary, McAfee, and Wolf, 1974; Bar-Yosef, 1987), Lisan Lake shrank drastically. During the dry climatic subphase of the Pottery Neolithic and Early Chalcolithic ages, the Dead Sea could have reached even below −400 MSL.

The warmest postglacial period is deduced as occurring between 7500 to 4500 years ago (Gary *et al.*, 1974). Neev and Emery (1995) interpreted a layer of rocksalt in the South Basin to represent the driest period of the Holocene, i.e. between 6500 and 5500 BP. Indications point to the destruction of Sodom having occurred somewhat later, about 2350 BC, and this would conclude the event to have taken place towards the end of Early Bronze within a climatic dry phase, thereby suggesting Dead Sea water levels would have been low.

The intense arid climate that persisted over prolonged periods of geological time would have resulted in conditions favourable for the development of playa formations. The mode of formation may be appreciated from a consideration of Figure 13.

Dead Sea Levels in Antiquity

The Book of Joshua (15:1-5) delineates the extent of the land given to the children of Judah:

> This then was the lot of the tribe of the children of Judah by their families; even to the borders of Edom the wilderness of Zin southward was the uttermost part of the south coast.

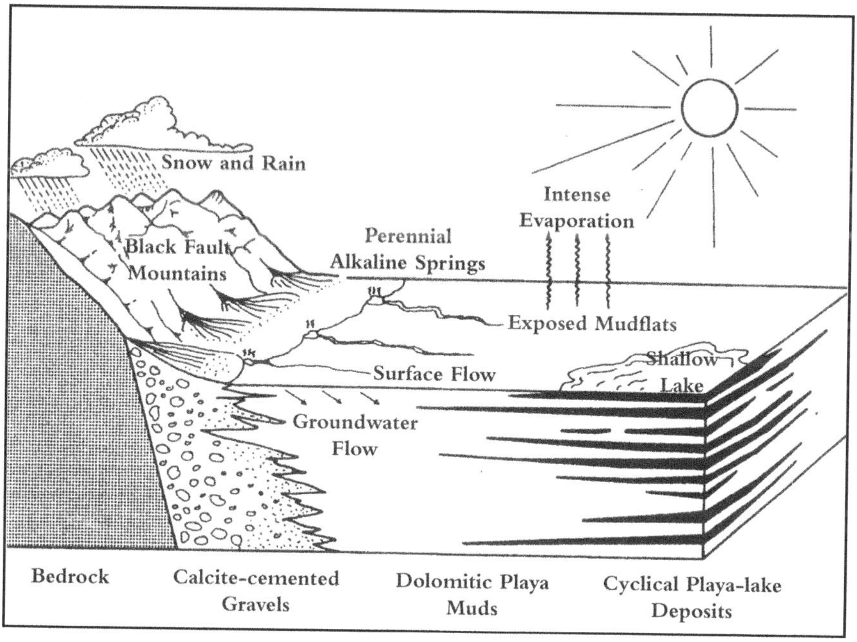

13. Playa formation within an intermontane basin (after Eugster and Hardie, 1978).

> And their south border was from the shore of the salt sea, from the bay that looketh southward:
>
> And it went out to the south side to Maaleh-acrabbim, and passed along to Zin, and ascended up on the south side unto Kadesh-barnea, and passed along to Hezron, and went up to Adar, and fetched a compass to Karkaa:
>
> From thence it passed toward Azmon, and went out to the river of Egypt; and the goings out of that coast were at the sea: this shall be your south coast.
>
> And the east border was the salt sea, even unto the end of Jordan. And their border in the north quarter was from the bay of the sea at the uppermost part of Jordan.

The inference is that the southern border of Judah extended from the south shore of the South Basin westwards to the Mediterranean, some of the place names given being associated with the wanderings of the Exodus. This would suggest that towards the end of the thirteenth century BC water levels in the Dead Sea were high enough to flood the South Basin. However, the context is not that clear.

Pliny the Elder visited the Dead Sea during the first century AD and the ratio he gives of width to length, measured in stadia, fit remarkably with the aspect ratio of the North Basin only. Thus, in the first century, it may be safely assumed that the South Basin was dry. It was still dry in Byzantine times as may be observed on the Madaba Map (AD 560), see Figure 15. Also, Strabo wrote:

> It [*the Dead Sea*] extends parallel to the coast to a length of slightly more than two hundred stadia, is deep to the very shore.

The Roman stadium is equal to about 185 metres, and from this it may be estimated that in Strabo's time (c. 54 BC to AD 21) the Dead Sea had a length of 37 kilometres, significantly less than in later periods of history. This length tallies approximately with the −430 metre contour of the North Basin, and from this it may be deduced that during the Roman occupation only the North Basin contained water.

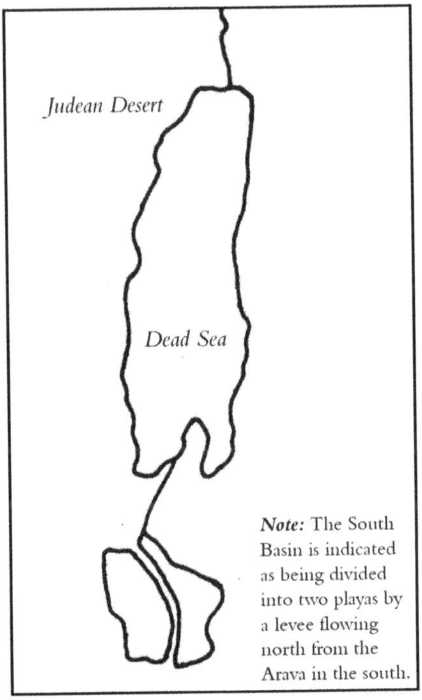

14. Detail of a map of Vespasian's campaign in Judaea showing the South Basin of the Dead Sea (after Hadas-Lebel, 1993).

Hadas-Lebel (1993) includes in her book on Flavius Josephus a map titled *Vespasian's Campaign in Judea* (late 67-June 69) – see Figure 14.

The provenance of this map is unknown, but it shows two bodies of water in the South Basin of the Dead Sea each isolated from the other by a levee running northwards from the Arava to the Lynch Strait. This suggests a low water level in the North Basin during the period, the strait being seriously diminished in width requiring, perhaps, only a shallow ford for its crossing. It is evident the levee was formed by sediments brought down the Arava in times of flood, thereby creating the two salt pans, or playas.

The Madaba Map

In 1894 a major archaeological discovery was made whilst excavating foundations for a new Greek Orthodox church at Madaba, Jordan. The previous church had been destroyed by earthquake in AD 746. The discovery proved to be a mosaic map of the entire region, now known as the Madaba Map, depicting the geography of the region from Lebanon in the north to the Nile delta in the south, with Jerusalem at its centre. Though the Dead Sea was included in the mosaic, only the North Basin was shown, the South Basin being omitted, see Figure 15.

Neev and Emery (1967) commented upon the Madaba Map in the following terms:

15. Detail from the Madaba Map (c. AD 560).

> The mosaic map of Madaba was constructed about AD 560 (Avi-Yonah, 1953). Although generally prepared on an inexact scale, it represents certain details quite precisely. The length/width ratio of the Dead Sea according to the map corresponds with that of Josephus, i.e. to the dimensions of the North Basin alone. . . . The similarity of the Madaba Map to the −435 metre contour, including the positions of river mouths, is striking.

The outline of the Madaba Map is superimposed on the shoreline of the Dead Sea (Figure 16). If the outline is accurate, and there is no reason to suggest otherwise, the waters of the Dead Sea during the period were confined to the North Basin. As Neev and Emery point out the configuration of the Madaba Map suggests the water level to coincide with the −435m contour. This is supported by the absence on the map of the tongue-like promontory of the Lisan Peninsula, and the embayment now constituting the Bay of Mazra'a. Such outstanding features are unlikely to have been overlooked by the makers of the map.

Neev and Emery (1995) also commented on the Madaba Map regarding ports on the Dead Sea linking the southern end with the north. They wrote:

> The east-west Roman road crossing Lynch Strait and the [*Lisan*] peninsula was first paved during the reign of Traianus (Trajan) Caesar in AD 106 as a branch of the north-south King's Highway along the Jordanian highland (Har-El, 1971). The strait was crossable then because the south shore of the sea was north of the peninsula. As Zoar of the 1st century AD was a seaport, it had to be on the shore and must have been north of that transverse road or near the head of the Bay of Mazra'a.

The depiction of vessels on the Madaba Map carrying cargo implies a seaborne trade on the Dead Sea, the most extensive of which would be trade linking the Lisan Peninsula with Jericho. Nissenbaum (1990) points out such commerce is known to have taken place since Hellenistic times, and probably for many centuries before, the Bay of Mazra'a being the only natural deepwater haven in the vicinity. During Byzantine times a small fortress existed, named al Karia, situated on a rise at the northern tip of the Lisan Peninsula. This would have overlooked the port itself, which is believed to have been called Zoar. Despite the similarity of names, the port of Zoar on the Lisan should not be assumed to represent Biblical Zoar to which Lot fled.

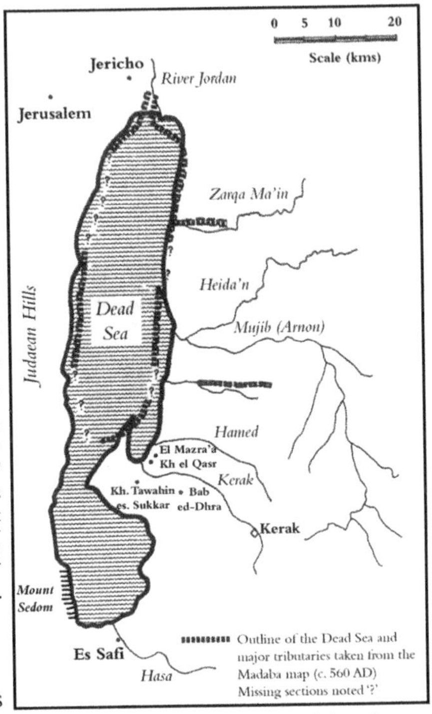

16. The outline of the Madaba Map superimposed on the Dead Sea (c. AD 560).

Biblical Zoar

The identification of Biblical Zoar has been a source of debate among scholars, for it was the only city mentioned in Genesis that reportedly escaped the destruction. Some have made a case for Zoar being identified with present-day Safi because it is now a thriving centre of population, being located on the fertile fan of Wadi Hasa at the southeast extremity of the South Basin, an existence that can be traced back to Antiquity. Additional supporting evidence comes from ruins of several water-driven mills in the area dating back to Mamluk, Crusader, and Muslim occupations. Though this points to the Safi area having been populated for at least a thousand years, and probably much longer, even extending back to Canaanite times, the South Basin of the Dead Sea would have constituted a salt playa for much of this period, and especially so at the time of the destruction about 2350 BC. Also, the Safi area is over 30 kilometres from the postulated Vale of Siddim which, in accordance with the thesis presented here, must be assumed as the location of Sodom. If any credence must be given to the flight of Lot and his daughters from Sodom, Safi was a long way to walk in one night. Furthermore, he and his family would have been in peril of being swept away in the ensuing tsunami following the loss of Sodom by earthquake liquefaction. Common sense suggests that in order to survive the disaster they must have fled uphill in accordance with the angels' exhortation to 'escape to the mountain, lest thou be consumed' (Genesis 19:17).

A far more attractive location than Safi for the site of Biblical Zoar lies uphill, within a night's walk from the postulated Vale of Siddim. Isaiah (15:5) contains a further clue:

> My heart shall cry out for Moab; his fugitives shall flee unto Zoar, an heifer of three years old: for by the mounting up of Luhith with weeping shall they go it up; for in the way of Horonaim they shall raise up a cry of destruction.

Neev and Emery (1967) claim a more accurate translation of this passage, and that the word 'fugitives' from the Hebrew, means 'bars', 'bolts', or 'latches', and that the translation of 'heifer' is wrong, and more correctly is a Moabite place name 'Eglat Shelishiya'. Luhith and Horonaim are Moabite place names in the Jordanian highlands, Luhith being a flat slab or board in Hebrew. They suggest the better interpretation of 'bars', 'bolts', or 'latches' from the foregoing is more suited to the description of the pass leading eastwards from the Lisan Peninsula, because of the highly tilted limestone strata (or *luhith*) flanking the road. This road, winding eastwards out of the Dead Sea valley, would have been a vital commercial, and strategic, link. It represents the shortest and easiest route from Judaea to Moab and further east.

The phrase 'latches of Moab' is particularly apt as the Eastern Border Fault region is very rugged. Furthermore, the subsidiary faulting at the western extremity of the pass creates a funnel-shaped flexure to the terrain resulting in a gentler topography at its lower elevations. Bab edh-Dhra would have constituted an important gateway to the pass. This archaeological site is located at an elevation of approximately −200 metres MSL occupying an area of about ten acres (in contrast Numeira is about two acres), and lies on the road leading from Mazra'a to Kerak just south of the Wadi Bab edh-Dhra, which drains into the Wadi Kerak some 500 metres or so to the north. The site is a distance of six to eight kilometres from the head of the present Gulf of Mazra'a considered to have previously constituted the Vale of Siddim. Further east of Bab edh-Dhra the road winds uphill, often through tortuous gorges. The road to Kerak is a unique natural feature, linking the Dead Sea with the hinterland of Moab.

At Bab edh-Dhra there are remains of fortifications and an abundance of graves. The latter indicates the site to have been settled for long periods in antiquity. In 1965-67 extensive excavations were carried out by P.W. Lapp on behalf of the American School of Oriental Research. After his death in 1970 work was continued by W.E. Rast and R.T. Schaub throughout 1975-81. The earliest settlement of the site has been dated to Early Bronze I (3150-3000 BC) with later settlement in Early Bronze II-III (3000-2300 BC). The date 2300 BC is considered significant as this is concluded as being some fifty years after the earthquake that destroyed Sodom and its satellite cities.

The Biblical description of the flight of Lot from Sodom is described in the following verses of Genesis:

> And it came to pass, when they had brought them forth abroad, that he [the angel] said, Escape for thy life; look not behind thee, neither stay thou in the plain; escape to the mountain, lest thou be consumed. (19:17)

> Behold now, this city is near to flee unto, and it is a little one: Oh, let me escape thither, (is it not a little one?) And my soul shall live. (19:20)
>
> Haste thee, escape thither; for I [the Lord God] cannot do any thing till thou be come thither. Therefore the name of the city was called Zoar. (19:22)
>
> The sun was risen upon the earth when Lot entered Zoar. (19:23)

Whether one accepts the Biblical account as truthful or with reservations, the conclusion must be that Lot fled uphill. He may not have made the mountain, but he managed to get to Zoar that night and it was spared. The site of Bab edh-Drha, being six to eight kilometres from the head of the Vale of Siddim, perhaps a few kilometres longer depending on the actual location of Sodom within the plain, could have been attained easily in a single night by the fleeing Lot accompanied by his daughters. The road to Zoar, or Bab edh-Drha, is the only road leading uphill out of the Vale of Siddim.

At an elevation of about 200 metres above the alluvial plain constituting the Vale of Siddim, Zoar may have suffered severe damage from the seismic forces unleashed. It would not, however, have suffered the same fate as the Vale of Siddim, as it is underlain by competent gravels and bedrock not susceptible to liquefaction. Also, it would have been at a high enough elevation to escape the effects of the ensuing tsunami. The archaeological site of Bab edh-Drha therefore must be favourably considered as representing Biblical Zoar.

The Plain of Salt

The run-off/evaporation ratios determined by Neev and Emery (1967), see Figure 12, would suggest that the South Basin constituted a salt playa for a period of at least three thousand years from the middle of the third millennium up to the Arab Conquest of AD 760, and the beginning of the collapse of the Byzantine Empire. This is supported by the implications of the Madaba Map (c. AD 560). Only later would water levels have risen to inundate the salt flats of the South Basin.

Before this, opposing armies could do battle upon the plain of salt, as evidenced by the following two Biblical narratives, which can be dated to approximately 1040 and 840 BC respectively:

> And David gat him a name when he returned from smiting of the Syrians in the valley of salt, being eighteen thousand men. (Samuel II, 8:13)
>
> He [Amaziah] slew of Edom in the valley of salt ten thousand, and took Selah [the rock] by war, and called the name of it Joktheel unto this day. (Kings II, 14:7)

The immensity of the slaughter, 18,000 and 10,000 men, reported in these two accounts, seems somewhat high for a society that was basically tribal in comparison to the present day. These two events occurred several hundred years before the Biblical accounts were written, and suggests the possibility of exaggeration in the numbers killed. Whether such an exaggeration was for priestly purposes, to demonstrate the power of the Israelites backed by the faith in their God to improve public morale, or some other purpose, will never be known. However, it is understood that a large number of graves have been unearthed in the Feifa and Safi areas at the south end of the Dead Sea, but whether they represent those who died by natural causes pursuing a pastoral but peaceful lifestyle, or killed in battle, has not been disclosed as far as the writer is aware.

In 1986 reports were made of the discovery of a cave complex near Safi. Located on a nearby mountainside at Deir 'Ain 'Abata, this has been proved to have been a Byzantine monastery dating back to at least the seventh century. The focal point is a triple apsed basilica built around a natural cave, with mosaic floors inscribed in Greek, dating to AD 606 and AD 691 respectively. Human remains exist, which have been identified as belonging to Early Bronze I (c. 3000 BC) and Middle Bronze II (c. 2000-1500 BC). The complex is presently referred to as the Sanctuary of Agios Lot, and attempts have been made to link the cave with the flight of Lot following the catastrophe that overwhelmed Sodom. The only acceptable conclusion, regarding the cave, is that it once served as human habitation, and thence as a sanctuary, religious or otherwise, prior to the Byzantine era.

Bitumen

The origin of bitumen, and its allied hydrocarbons, has long been a source of debate among petroleum chemists and geologists. When the petroleum industry was in its infancy there was a general belief that hydrocarbons had an inorganic source deep within the Earth's crust. Then the argument shifted in favour of marine, or terrestrial, sources originating with the decay process of organic remains. Possibly both are correct, for the bitumen deposits of the Dead Sea may be encountered both at the surface and at great depth. At the surface they occur as veins in hard rocks, cementitious material with gravels and other coarse sediments, and infillings to vugs, e.g. cavities. The most impressive source, however, is that which occurs in the form of floating blocks weighing several tons. The latter possess a remarkably high purity.

The sources of the bitumen are somewhat contentious. One claim suggests it to be from thermally immature bituminous chalks of the Upper Cretaceous in the centre of the Rift Valley, possibly beneath the Dead Sea itself. The current concept, however, is that the more fluid hydrocarbons arise from thermal cracking of heavier bituminous materials contained within source rocks at great depth. As a consequence the resulting lighter fractions migrate upwards along fractures, such as fault planes or other structural discontinuities in the host rocks, until they are eventually vented. Dislodged by periodic seismicity, the bitumen solidifies before floating upwards to emerge at sea level in various forms as will be discussed in Chapter 4.

As the source of the bitumen remains uncertain, it is possible there are several points of emergence on the sea floor. Though the current thinking is that the source rocks are likely Cretaceous oil-shales, as these outcrop widely throughout the region, rocks of Jurassic age are present at greater depth. The Jurassic, like the Cretaceous, was a period noted for abundant fossil remains and, consequently, such an origin should not be dismissed too readily.

Dead Sea Salts

The Dead Sea is renowned for its high concentration of dissolved salts. As a result the water has such a high density that, as Strabo noted two thousand years ago, it is virtually impossible for a human being to drown. Numerous measurements have been made of the specific gravity of Dead Sea water, but the variation is highly dependent on many factors rendering them of localised value only. The overall range is generally considered as between about 1.16 and 1.24, according to Braslavy (1956).

In their detailed investigations into the chemical and physical characteristics of Dead Sea water Neev and Emery (1967) identified many of the compounds and the variation in their ionic distributions. This work identified a southerly current flow through the Lynch Strait into the South Basin, with an increasing concentration through evaporation leading to deposition of magnesium, potassium, bromine and sodium salts. They also showed that in general the salt content of the water increases with depth. There is, therefore, considerable variation throughout the Dead Sea in the distribution of dissolved chemical substances, not only between localities, but also as regards depth. Table A lists the mean concentrations of the more prevalent ions. From these the average salinity (as a sum of the ionic concentrations) exceeds 322 grams per litre resulting in an overall water density of 1.23 grams per cubic centimetre.

Sauer (1978) reported that the dissolved-solids concentration of the Dead Sea approximated 322,000 milligrams per litre and, that at this concentration the water is virtually saturated resulting in continuous precipitation of certain salts. This precipitation is aided by evaporation which amounted (in 1978) to 1,310 millimetres per annum. The natural precipitation of salts from Dead Sea water, as a consequence of evaporation, is influenced by various environmental factors, including annual fluctuations in climate, e.g. temperature, rainfall, etc.

One of the more spectacular consequences of the intense period of evaporation, during which the water level of Lake Lisan fell dramatically, was the deposition of a rhythmic sequence of banded gypsum and aragonite. Remnants of such sequences of deposition may be observed in the flanks of the Dead Sea valley where they have not been removed by subsequent erosion. The alternating laminae reflect annual depositions of the minerals involved. The darker laminae, composed mainly of detrital matter but with a relatively

17. Layered deposits of gypsum and aragonite in a road cutting.

high gypsum ($CaSO_4.2H_2O$) content, are considered to represent winter depositions, whereas the whiter lamina, composed of aragonite ($CaCO_3$), would have precipitated during warmer periods. Aragonite is a form of limestone and, changes its crystalline form over time to evolve into calcite.

Figure 17 illustrates a section in a road cutting through the banded gypsum/aragonite. There is a basal horizontally bedded formation overlain by a continuation of the sequence, but one in which the laminae have become highly contorted. Evident also is a well-defined shear zone (right of centre). It is obvious that the upper materials have been disturbed, probably by earthquake, whereas the lower have resisted deformation.

The Sedom Clay

More recent sediments can be identified as being laid down over the last 20,000 years following an abrupt climate change from pluvial to dry (see Figure 12). A major component stratum at relatively shallow depth beneath Dead Sea water levels has been termed the Sedom Clay. Soil exploration drilling has shown this to be in excess of forty metres thick towards the extreme southwest corner of the South Basin. Because of its nature, in being a chemical precipitate, it may be assumed to be pervasive among recent Dead Sea sediments, and blanket older sediments.

The term 'clay' as applied to the Sedom Clay, however, is somewhat of a misnomer as the clay sizes are comprised of particulate calcite amounting to over 90 percent of the constituent particulates, the remainder consisting of gypsum, detrital quartz, limestone fragments, salt and traces of clay minerals. From a behavioural standpoint it might more properly be described as a silt or clayey silt. The presence of oxidised minerals located at depth underlying unoxidised minerals at higher elevation indicate the Sedom Clay to have been laid down during at least two separate depositions. These are likely to have been preceding and following prolonged periods of high rainfall/low evaporation.

In geologic terms the Sedom Clay may be considered to fall into the category of a micritic limestone which, after the further elapse of geologic time and consolidation from an accumulation of further sediments, would become limestone in the more accepted meaning of the term. At the present time the Sedom Clay is a soft soil of relatively low strength possessing a sensitivity of about 3 (sensitivity is defined as the ratio of undisturbed strength to remoulded strength of clay). Consequently, Sedom Clay is susceptible to liquefaction when subjected to seismic forces above a certain magnitude, especially as the clay contains laminae of silt and fine sand. The potential liquefaction of this stratum is discussed in greater detail in Chapter 6, as its likely existence beneath the overlying alluvium underlying the Vale of Siddim would have made that area vulnerable to greater liquefaction than otherwise and, thereby, have contributed to its loss beneath the waters of the Dead Sea.

Lot's Wife

The Dead Sea region is noted, amongst other things, for the tourist attraction known as Lot's Wife. The narrative of Genesis describes in vivid detail how she was turned into a pillar of salt for her disobedience in looking back upon

18. The pinnacle known as Lot's Wife, Jordan, © Sergio Ponomarev.

the destruction of Sodom. However, depending on which side of the Dead Sea the tourist finds himself dictates which Lot's Wife he will be shown by his guide, for there are two pillars answering, somewhat contrarily, to that name on opposing sides of the Dead Sea. The one on the Jordanian side is near to Deir 'Ain 'Abata, north of Es Safi, close to a cave reputed to be the one used by Lot and his daughters following their flight from Sodom, see Figure 18. The one on

19. The pillar known as Lot's Wife on Mount Sedom, Israel.

the Israeli side stands on the skirts of Mount Sedom (the Arabic name for which is Jebel Usdum) and overlooks the South Basin, see Figure 19. Because of the soluble nature, and consequent ease of erosion of the rocks concerned, the shape and size of the pillars have a disconcerting habit of changing with time.

The first recorded observations relating to Lot's Wife on Mount Sedom were made during the United States Navy Expedition to the Dead Sea (1847-48). Lynch (1849) and Montague (1849) write as follows:

> [W]e saw on the eastern side of Usdum, one third the distance from its north extreme, a lofty, round pillar, standing apparently detached from the general mass, at the head of a deep, narrow and abrupt chasm. . . . We found the pillar to be of solid salt, capped with carbonate of lime, cylindrical in front and pyramidal behind. The upper or rounded part is about forty feet high, resting on a kind of oval pedestal, from forty to sixty feet above the level of the sea. (Lynch)

> On pulling round the shores of the sea, we saw an immense column, rounded and turret-shaped, facing towards the southeast. This, we were told by our Arabs, was the Pillar of Salt, in which Lot's wife was encased at the overthrow of Sodom. . . . It was measured and found to be sixty feet in height, and forty-five feet in circumference. We cannot suppose Lot's wife was a person so large that her dimensions equalled those of this column. (Montague)

Geologically, Mount Sedom is known as a diapir, i.e. a vertical intrusion of relatively mobile material into, or through, overlying rocks piercing them from below. They are associated with weaknesses, or structural deformities in the Earth's crust. Mount Sedom represents an immense diapir, but it is not the only one in the vicinity, see Figure 9. It forms a ridge eleven kilometres long and one and a half kilometres wide, rising to an elevation of −160 metres, i.e. about 250 metres above the 1992 level of the Dead Sea. The depth of its base is considered to be about three kilometres below surface (Gardosh *et al.*, 1997).

Frumkin and Ford (1995) report the existence of over one hundred caves within Mount Sedom of sufficient size to be physically entered, and one system of galleries to extend a total length of 5700 metres. These have been produced by solution of the salt by freshwater penetrating fissures and crevices within the salt mass. Many of the caves are in the process of being actively enlarged, but others are of a remnant nature. Both active and inactive systems present valuable information relating to the historic variation of external lake levels.

Pillars of salt are unique morphological features, being found only in arid climates due to their extreme solubility from atmospheric waters, i.e. rain and surface runoff. The harsh dry climate of the Dead Sea, however, is very favourable for their preservation. Frumkin (2009) has shown the twenty metre high pillar of salt known as Lot's Wife, on Mount Sedom, to be a remnant of cave collapse, the caves collapsing either side of the pillar, presumably, as a consequence of seismic activity. The caves, which are karst in nature, originated as solution features from downward migration of fresh water. Frumkin (1997) reports the caves have yielded thirty-three specimens of wood embedded in cave floor sediments, and these specimens have been dated by radiocarbon techniques giving an overall range in age from 6000 BC to AD 1800. The existence of these caves, and their location within the mass of Mount Sedom, has enabled useful information to be deduced respecting levels at which the Dead Sea stabilised for any length of time. Such studies are valuable in helping to resurrect Dead Sea water levels. Implicit in this is the fact that caves existed within Mount Sedom during the period when the Sodomites were active in the region. It is possible, therefore, that some of the outer workings may constitute adits used for mining rock salt, though no archaeological evidence, in the form of recovered artifacts, is understood to have been reported.

Ancillary geological studies have indicated the Mount Sedom diapir to be active and currently growing by 9.3 ± 3.5 millimetres per year. Assuming the current rate of growth represents the average rate of growth since the demise of Sodom, it is calculated that the pillar of salt known as Lot's Wife has risen some 40 metres over the last four millennia. The assumption of a constant rate of growth over such a long period of time, however, is unlikely to reflect reality.

It is evident from the geological evidence that neither of the two 'pillars', currently touted as Lot's Wife, may be considered as representing that unfortunate woman. Another explanation must be found to give credence to the Biblical narrative, and this is given in Chapter 8.

20. The Bay of Mazra'a, previously the Vale of Siddim, viewed from the north.

Vale of Siddim

If, as Strabo states, 'Sodom was the metropolis', and the exploitation of bitumen resources was the *raison d'être* for its existence, a supporting agricultural base together with the assurance of a reliable supply of fresh water, were essential prerequisites to support a community of any size, especially one calling itself a 'city'. As we are told in Genesis (Ch. 12), after Lot and Abraham parted, the former took his herds and 'pitched his tent towards Sodom'. These three factors were all present within the postulated Vale of Siddim before the vale was lost through earthquake and inundated by the sea. The alluvial plain formed by the sediments brought down from the Jordanian hills by Wadi Kerak and its tributaries, and other contributory streams, would have been well-watered and fertile, probably the most extensive and luxuriant of those towards the southern extremity of the Dead Sea basin.

The alluvial plain so formed over the preceding millennia would have created an area of approximately sixty square kilometres, with a shoreline approaching eight kilometres long. That shoreline would have advanced annually; being fed by the sediments brought down by the streams into the Vale of Siddim, and would have encroached progressively into deeper and deeper waters. It may be imagined that the ephemeral waters of the Wadi Kerak meandered through this vale, possibly constrained by canal works to serve the purposes of the Sodomites. Following the loss of the Vale of Siddim by earthquake induced liquefaction as postulated (Harris and Beardow, 1995), the shoreline would have been substantially modified resulting in the formation of the present embayment between the Lisan Peninsula and the eastern shore, now known as the Bay of Mazra'a. Figure 20 shows the

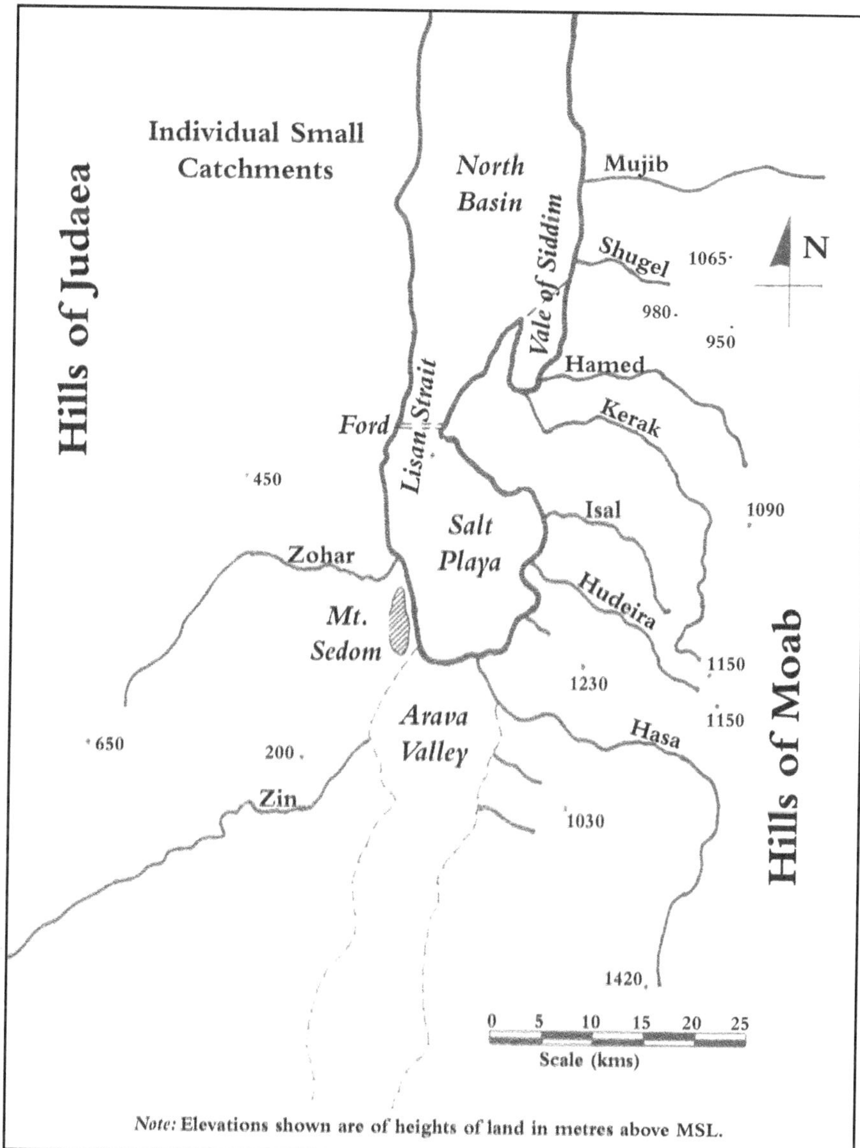

21. Watercourses feeding the southern part of the Dead Sea.

present bay viewed from the north, it would have presented a far different sight when it constituted the Vale of Siddim before the destruction.

Bentor (1989), in reference to the battle of the kings in the Vale of Siddim, claims the word 'siddim' means 'lime'. Much of the western shoreline of the Dead Sea, and the Lisan Peninsula, is comprised of soft, white, Late Pleistocene beds consisting essentially of glaringly white layers of aragonite and gypsiferous marls. The Vale of Siddim thus formed by the alluvial plain of the Wadi Kerak prior to its loss by liquefaction, would have been flanked by the Lisan Peninsula to the west, and the rising slope of the eastern shore composed essentially of older, harder, and more resistant limestones of Cretaceous age. The Vale of Siddim, or Valley of Lime, would have been an apt name for such a place.

Water Supply to Sodom and the Vale of Siddim

Figure 21 shows the main watercourses discharging into the southeastern quadrant of the Dead Sea, of which several flow directly into the Vale of Siddim. These streams are, for the most part, ephemeral, and rain can rapidly transform dry stream beds into raging torrents. Figure 22 depicts the dry floor of a wadi swept clean of alluvial material except for boulders, a testimony to the force of the flow when in flood. The traveller should be aware of this danger.

Watercourse	*Catchment area (Sq. kms)*	*Length (kms)*	*Elevation at source (m above MSL)*	*Remarks*
Shuqei	80	15	950	Discharge into Vale of Siddim.
Hamed	150	25	1090	Discharge into Vale of Siddim.
Kerak	170	40	1150	Discharge into Vale of Siddim.
Isal	140	25	1150	
Hudeira	120	30	1150	

Table B – Hydrological data of wadis discharging into the Vale of Siddim and vicinity.

The main hydrological characteristics of watercourses feeding the Vale of Siddim and its immediate vicinity are tabulated in Table B. It will be noted that Wadi Kerak is the largest contributor of fresh water to the Vale of Siddim, its catchment area totalling about 170 square kilometres. The combined catchments for Wadis Shuqei, Hamed and Kerak, all of which flow into the Vale of Siddim, total 400 square kilometres. Assuming an annual precipitation of about 30 centimetres, with a run-off coefficient of 0.3, and with evaporation and other losses of about 50 percent, the average total annual water supply from these sources can be estimated as being of the order of 20 million cubic metres. This is a lot of fresh water, which if stored and controlled efficiently could have supported a large population engaged on a multitude of industrial tasks. The vale was, therefore, well favoured.

During the period when Sodom was prosperous and flourishing the sources of water would have been much as they are now, but likely of lower seasonal fluctuation due to an abundance of Mediterranean mountain flora clothing the

22. The rock floor of a wadi swept clean by torrents.

Jordanian hillsides. These forested slopes would have mitigated the erosion in times of heavy rainfall, leading to greater groundwater retention in the soils and enhanced grazing and agriculture. It is reasonable to assume that the area around Sodom would have presented a well-watered aspect, one pleasing to the eye. According to Reifenberg (1955) the arid semi-desert environment of today is the consequence of degeneration following the Arab conquest.

Given the region around present-day Bab edh-Drhra and El Mazra'a was an agricultural centre, famous for its 'fruits of the palm' as late as the twelfth century, if we can believe Foulcher of Chartres, a priest accompanying the Crusader raid into Moab in AD 1100 (Lamb, 1953), then the Lisan required a harbour. The existence of such a harbour is supported by the twelfth century Arab geographer Al-Idrisi, who tells of the dates of Zoar being shipped to Jericho by boat (Braslavy, 1956). The only haven suitable for shipping would have been on the Lisan Peninsula.

It would seem that even more recently there is ample evidence of fruitfulness in selected locations around the fringes of the Dead Sea, both in the North and South Basins. Lynch (1849) observed both standing and fallen timber at the base of cliffs on the Lisan Peninsula, and Kyle (1928) observed this about the South Basin:

> There were trees little and big, from mere saplings to trees a foot in diameter. It was a ghost of a forest once living and green, now dead and blasted a ghostly white by the salt. In places the submerged forest was seen to extend out a mile from the shore on the eastern side and somewhat less from the western side of this narrow arm of the Sea in the southern end.

Numerous water courses, generally of an ephemeral nature, are distributed around the perimeter of the Dead Sea, and these have given rise in recent years to a resurgence of intensive agriculture in oasis-like environments. The fertility of the region, therefore, is not to be doubted when water supply is abundant. Neither is the evidence, now encrusted in salt, that such fertility once extended offshore before being overwhelmed by the rising waters of the Dead Sea.

Reconstruction of Dead Sea Water Levels

Inland lakes, such as the Dead Sea, provide a unique opportunity for the geologist to reconstruct lake level history, as only two factors are of major impact, precipitation and evaporation, and their relative balance. Normally, there are no external influences, such as that of sea level, to confuse the evidence. Several approaches may be adopted: (a) sedimentological studies which provide clues of a lithological, mineralogical, palaeontological and palynological nature; (b) studies of a geomorphological nature based on evidence of ancient beach terraces, and other physical changes to the existing land forms; and (c) historical studies relating to archaeological sites and encampments of a cultural nature. All these specialist spheres of knowledge contribute to a better understanding of lake level movements throughout the millennia.

The Dead Sea is isolated from general world sea levels by a height of land along the connecting valley with the Gulf of Aqaba (the Arava graben), and has been so isolated for several millennia. During Pleistocene and Recent times the region experienced intense precipitation following which there has been a steady, but irregular diminishment of water level. This has been discussed earlier in this chapter. The lowest level of the Dead Sea in historical times seems to have been attained in the period 2000-1000 BC, i.e. about −460 metres MSL. This would have totally exposed the floor of the South Basin and created a wide fringe of beach around the shores of the North Basin, the width varying with immediate offshore water depth.

Frumkin *et al.* (1991) gave evidence from seismic reflection data that revealed the existence of relict canyons and buried channels at the mouth of the Jordan River suggesting that Dead Sea levels were below −400m MSL for much of the Holocene, a geological period generally considered to have commenced about 12,000 years ago and extending to the present day. Neev and Emery (1967) reported that during this period the South Basin of the Dead Sea has subsided at a rate of 1-10 metres every thousand years as a result of slippage along faults bounding the basin. This subsidence has resulted in a tilting of the Mount Sedom diapir towards the east. From a geological standpoint the region may be considered extremely active.

Yechieli *et al.* (1993) reported the findings of boreholes put down in the alluvial fan of Nahal Ze'elim, which recovered wood fragments. These could be radiocarbon dated to 8255 years, 8390 years, 8440 years, and 11,315 years BP respectively. As a result salt beds deposited after 11,315 years, but before 8440 years BP were concluded as having been formed at a low water level in the transitional period between Lake Lisan and the present Dead Sea.

According to Frumkin (1997) at least ten stages of water level can be identified, and these are reproduced in Table C. The last major high level period is concluded as having occurred in the interval 4200 to 2300 BC, following which major desiccation took place, there being a good correlation between the lake level curve, sediments and pollen counts. During Frumkin's Stage 3 (4200-2300 BC), a period of almost two thousand years, the Dead Sea basin would have witnessed the likely blossoming of Canaanite civilisation. Urban centres, such as Sodom and Gomorrah, would have developed, grown, and flourished during the Early Bronze age up to the end of EBIII.

Period (years)	*Level (metres) MSL*
10, 000 to 5800 BC	−400
5800 to 4200 BC	−404
4200 to 2300 BC	−359 to −355/−389 to −345
3200 to 1500 BC	−380
1500 to 1200 BC	−393 to −363
1200 to 100 BC	−400 to −395
100 BC to AD 400	−387 to −375
AD 400 to AD 900	−436 to −390
AD 900 to AD 1300	−382
AD 1300 to Present	−390 to −400 (to 1980)
	−400 (fourteenth century)
	−402 (early nineteenth century)
	−391 (end nineteenth century)
	−418 (2004-2005)

Table C – Stages of Dead Sea water levels as identified by Frumkin (1997).

Frumkin tenders a considerable amount of evidence, some based on radiocarbon dating, others based on morphological reasoning, to conclude Dead Sea lake levels were relatively stable during the period 4200 to 2300 BC ranging mostly between −359 and −355m MSL. This means the South Basin would have been flooded, with water depths amounting to some forty metres. It is unlikely, with this depth of water that the South Basin would have dried out to become a salt playa. Though the Lisan Peninsula would not have been submerged, the Lynch Strait would have been about five kilometres wide. A wide range of paleoclimatic evidence in various parts of Israel and the Near East suggests the climate was moist, and the Dead Sea attained its highest level during the last 7000 years.

During Frumkin's Stage 4 (2300-1500 BC) lake levels began to recede, and the South Basin dried out completely becoming the salt playa suggested by Samuel II (8:13) and Kings II (14:7). Therefore, it is not unreasonable to suggest that at the time of Sodom's destruction water levels in the Dead Sea, were lower than about −400m MSL. In other words the South Basin contained no appreciable depth of water. However, Frumkin's reconstruction of Dead Sea levels is based largely upon radiocarbon dating of organic debris found in caves on Mount Sedom, and elsewhere in the area. As will be pointed out later such techniques are not without shortcomings, and are often found to be of dubious reliability (see Appendix I).

There is nothing to contest Frumkin's conclusions based upon the observations made. However, no allowance appears to have been made for any possible tsunami effects to which the Dead Sea may have been subject. As will be discussed later, the destruction of Sodom by earthquake liquefaction would have resulted in a massive tsunami event, one which would have cast flotsam and debris many metres higher than existing lake levels. Subsequent radiocarbon dating of such organic debris, though correct in itself would, nevertheless, result in an invalid conclusion, suggesting lake levels to be higher than they actually were.

Neev and Emery (1995) report the following:

> The climatic fluctuation pattern for Holocene identified from gamma-ray logs of drill holes in the south basin agrees with that of intensities during wet subphases inferred from the Mount Sedom caves. . . . According to radiocarbon analyses of tree twigs found in galleries of the caves, the peak of that phase should have been between 4580 +/− 90 BP and 4250 +/− 95 BP – between the end of Early Bronze II and the middle of the Intermediate Bronze age. Traditional archaeological chronology of Early Bronze sequences at the Bab edh-Dhr'a and Numeira sites, the Beth She'an basin, and Beth Yerah clearly indicates that this wet climatic phase occurred sometime after the end of the Intermediate Bronze age and ended with the Late Bronze between 3900 BP and 3200 BP. A discrepancy of 600 and 1,000 years between apparent ages of this wet phase is inferred by comparing results of the geochronological-radiocarbon dating analyses made in studies along the Dead Sea-Jordan valley at archaeological sites near the geological ones. Such a gap is too large to have been caused by errors in radiocarbon analysis. An alternative is an earlier wet subphase of appreciably less intensity and shorter duration (200 to 400 years) sometime between 5000 BP and 4600 BP during Early Bronze I and II. . . . This interpretation disagrees with radiocarbon ages of the Sedom cave driftwood. *To avoid confusion, traditional archaeological chronology has been used as far as possible in correlation efforts, not necessarily because it should be considered correct but because it is more consistent and based upon a broad spectrum of regional data as well as on a large number of radiocarbon analyses* [author's italics].

The problem with some of the foregoing inconsistencies is that an over-reliance is often attributed to radiocarbon dating methods. The popular impression of radiocarbon dating is that it is an exact science. Nothing could be further from the truth, as much relies on the accuracy of the basic assumptions made, as well as the techniques and procedures used. As a result comparison is required (if possible) with other methods to ensure confidence in the end result. That is why the statement by Neev and Emery reproduced above has been italicised. Appendix I contains a brief summary of radiocarbon dating with some of the pitfalls leading to inaccurate conclusions.

As a demonstration of some of the anomalous results that can occur from carbon-dating Neev and Emery (1995) describe a one metre long wooden stick, three centimetres in diameter, discovered in an Early Bronze I chamber tomb at Bab edh Dhra. When subjected to radiocarbon dating the two extremities yielded ages of about 3100 BC at one end and zero at the other. There is a perfectly good explanation for anomalies like these, but they can only be given when the situation is fully understood, and the limitations of testing known. To indicate the scale of confusion surrounding radiocarbon testing one has only to look at the controversy regarding the Shroud of Turin.

The Shroud of Turin

The origin of the Shroud of Turin continues to be the subject of intense, and often passionate, rhetoric among scientists, theologians, and historians. In 1988 a sample of cloth was removed from the shroud, and dated independently, using state of the art radiocarbon testing techniques, by three highly respected laboratories. They were Oxford University, the University of Arizona, and the Swiss Federal Institute of Technology. The results were:

Oxford	646 ± 31 yrs BP
Tucson	750 ± 30 yrs BP
Zurich	676 ± 24 yrs BP

These results were subjected to statistical analysis, from which it was concluded the origin of the shroud could be dated to between 1273 and 1288 (with 68 percent confidence) and between 1262 and 1384 (with 95 percent confidence). However, despite the seemingly reasonable agreement of the tests, the results only provoked a storm of criticism, with charges the shroud had been contaminated by handling, by bacteria, by pigmentation, or the findings skewed in the pre-treatment process. All these criticisms appear fair and, if accepted, would have resulted in a more recent age being assigned to the shroud. A procedure that had been undertaken to settle an age-old enigma once and for all, in the end failed to please anyone.

Post-Destruction Occupation

Reifenberg (1955) refers to the verse in Chronicles I (27:28) as a confirmation that the Dead Sea valley was undoubtedly wooded, including groves of oaks, pines, cypresses and sycamores, sycamores being planted for timber. He writes that:

> It is known that forty ships brought cedars from Mount Lebanon, or Mount Amanus in Syria, to Egypt during the reign of Snefru as early as the third millennium BC. Thutmoses III in the fifteenth Century BC boasted that he had cut cedars in Lebanon of sixty cubits [*almost 90 feet*] in length.
>
> Syria, formerly known as one of the Roman 'granaries', is an outstanding example of the decline of agriculture and civilization caused by soil erosion. The destruction of vegetation is most devastating in the mountain region, where steep slopes are still intensively cultivated. In the pine forests the fallen needles are collected for fuel, and the formation of a natural protective humus cover thereby prevented.

Reifenberg claims that the population of the Negeb region during Byzantine times was 80-100,000, but during the final years of the British Mandate had shrunk to about 50,000 as a consequence of desertification resulting from soil impoverishment. Though the production of local timber in Byzantine times was adequate, the size and quality was diminishing as deforestation continued for charcoal production. The availability of large sized timbers was severely limited, and during the Siege of Jerusalem (June-July 1099), the siege was halted while the Crusaders awaited a shipment of timber to arrive by sea. Even this imported timber had to be supplemented by wood torn from the Genoese galleys that brought it.

3. The Canaanites

Archaeology

The Canaanites are a somewhat elusive people, considered to have occupied a vast region of the Near East, including the entire Mediterranean coast extending inland into parts of modern-day Syria and Jordan. The earliest known textual reference to Canaan is on a statue found during excavations in 1939 at Tell Atshanah in ancient Alalakh (Northern Syria). The statue depicts Idri-mi, the ruler about 1550 BC. The inscription on the statue in cuneiform script refers to Canaan as *ma-at ki-in-a-nimki* (or land of Canaan). In the Old Testament Canaan is known as *kĕna'an*, or *'eres- kĕna'an* (Mitchell, 1988). Unlike the Assyrians and Egyptians, who flanked them to north and south, the Canaanites left few written records. Those which have survived are mainly in the form of commercial records inscribed on clay tablets. An example of typical cuneiform script is depicted on Figure 23. Our knowledge of Canaanite civilisation has been acquired through the painstaking efforts of archaeologists who, though deprived of contemporary references, have nevertheless brought about a degree of understanding of their culture, and way of life by means of excavation. Tubb (1998) sums up the situation admirably:

> For the Canaanites, unlike the Egyptians, Babylonians, Assyrians or Hittites, did not leave to posterity wealthy archives of contemporary texts. In the case of these other peoples, such texts have enabled historians to flesh out their ancient lives, as preserved by the purely material remains, with vivid details of their social institutions, international relations, commercial dealings, religions and beliefs, literary traditions, mythologies and a wide range of aspects of their daily lives.

The Canaanites were a Semitic race akin to the Israelites, a thesis often brought forward. Canaanites taken into captivity might well be allied to ancestors who had been forced to abandon their homes in Canaan following the destruction of Sodom in c. 2350 BC, and had sought refuge in Egypt. Scholarship on such matters is understandably vague. However, as the destruction of Sodom predates the Exodus, the Sodomites were obviously Canaanites.

Occupation of the land of Canaan can be traced back several thousands of years into the dimness of prehistory. At Jericho excavations have revealed

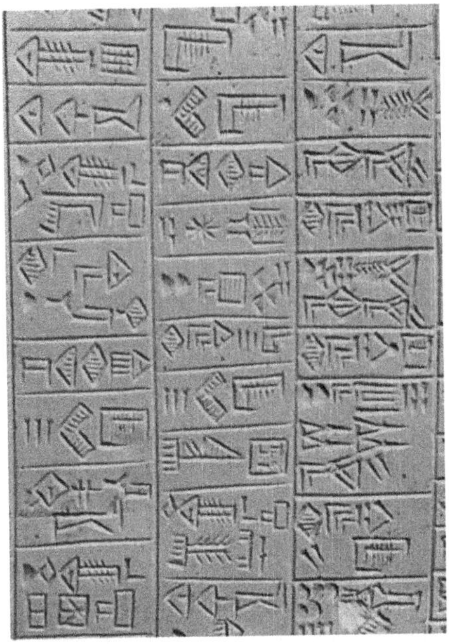

23. Sumerian cuneiform script c. 2600 BC (US Public Domain).

a sequence of occupations extending into the Neolithic Period (8500-4500 BC). Remains from the pre-pottery period (7200-6500 BC) have been uncovered also at Beidha, Jordan. The subsequent slow but steady advance in Canaanite culture and civilisation from these early origins can be traced through studies of pottery, statuettes, ornaments and other artifacts discovered within the acknowledged bounds of Canaan. It may be concluded the Dead Sea valley has been occupied by mankind for over ten millennia.

The Paleolithic, or Early Stone Age, was a period when stone implements and weapons were crudest, and man was nomadic. The Neolithic, or Late Stone Age, in contrast was a period when stone implements began to be polished, animals domesticated, cereals cultivated, cloth woven and religion structured. Of particular interest is the fact that bitumen, identified as from the Dead Sea region, began to be used for lining reed baskets and fixing flint axe heads.

The ensuing period, the Chalcolithic, is so termed as it is derived from the Greek *chalkos* (copper) and may be perceived as a transition between the Neolithic and the subsequent Bronze Age when tools, weapons and various utensils began to be fabricated out of copper and copper based alloys. The Judaean Hills abound with chert strata. The flint available would have yielded a rich source of material for the manufacture of stone tools, and the mines of Sinai and the Negev would have produced copper. Copper ores are relatively easy to smelt, and a flourishing metal-working industry must have existed as numerous copper objects have been discovered dating back to the Chalcolithic. Therefore, there is no reason to suggest that Canaan, during the Early Bronze age, was not at the forefront of technological progress. In fact the Dead Sea valley, as will be argued later, might well be perceived as a hive of industry with its close ties to Egyptian commerce and technology.

Forbes and Dijksterhuis (1963) make the point that tools shaped from stone and flint were not immediately supplanted by tools made of metal. Copper tools, even those alloyed with tin, were not much more effective than stone in many ways. They were also more expensive to produce and, generally speaking, were the preserve of the wealthier classes. They write this of Stone Age man:

By directing his chipping of flint towards the production of useful flakes he produced different types of knives, saws, and chisels which allowed him to fleece and skin animals of the steppe and furnished him with useful weapons for the chase. Nor should we underestimate the usefulness of such tools. Modern experts, comparing the cutting angle and shape of such prehistoric tools with modern ones, find that they do not differ in principle and that many a prehistoric tool would be as efficient and economic as a modern tool if guided by a skilled hand. In the New Stone Age man learned to polish harder stones, such as granite, and developed further shapes of useful stone tools to complete a tool chest which would have compared favourably with that of any good craftsman of a century ago, before the advent of modern machinery.

It was during the Bronze Age (3100-1200 BC) that sufficiently rapid cultural advances were made enabling the period to be subdivided into Early Bronze (3100-2000 BC), Middle Bronze (2000-1750 BC) and Late Bronze (1750-1200 BC) with appropriate transitions between these subdivisions. The Bronze Age was in turn followed by the Iron Age after man learned the more difficult art of smelting and working iron about 1500 BC.

24. Early Bronze Age Pottery from Khirbet Kerak (Courtesy Hanay).

25. A Mesopotamian cylinder seal (Courtesy Jastrow).

It is in the Chalcolithic and Early Bronze that decorative objects began to make their appearance in increasing numbers, reflecting the initiation of an artistic tradition. Representations of male and female figures, animals, and natural objects are carved in bone or ivory, wall-paintings are made exhibiting elaborate geometric designs, and bitumen makes its appearance in jewellery and ornamental art. There are also cultural changes in disposing of the dead for their remains now begin to be buried in ossuaries made of stone or earthenware. Figure 24 shows the type of pottery in general use during the period recovered from a tell on the south shore of the Sea of Galilee, and Figure 25 is an example of a cylinder seal from Mesopotamia used to inscribe clay tablets.

Under the supervision of Jonathan Tubb, archaeologists from the British Museum have focussed on two sites in the Jordan River valley, both within present-day Jordan, and have established the existence of two important centres of Canaanite commerce. These are Tiwal esh-Sharqi, where excavations began in 1983, and Tell es-Sa'idiyah, considered to be the Biblical city of Zaretan, where excavation started in 1985. Zaretan is on the east bank of the Jordan River where the Israelite army, commanded by Joshua, crossed into the Promised Land:

> That the waters which came down from above stood and rose up upon an heap very far from the city Adam, that is beside Zaretan: and those that came down toward the sea of the plain, even the salt sea, failed, and were cut off: and the people passed over right against Jericho.
>
> And the priests that bare the ark of the covenant of the Lord stood firm on dry ground in the midst of Jordan, and all the Israelites passed over on dry ground, until all the people were passed clean over Jordan. (Joshua 3:16-17)

The parting of the Jordan River, to permit Joshua and his host to pass dry shod, has been interpreted as evidence of a landslide temporarily blocking the river. Given the seismicity of the region, such an explanation has plausibility. Further evidence of this physical explanation is reported by Bentor (1989):

> It is here that the river crosses the major strike-slip, the Jordan Valley Fault; its bed is very narrow here and deeply incised between almost vertical walls cut into the unconsolidated lacustrine chalks of the Late Pleistocene Lisan Formation. In the earthquake of AD 1546 the banks of the Jordan River caved in again at Damiya and the water of the river was cut off for two days (Braslavski, 1938); and it happened again at the same locality in 1267 (Watson, 1895), in 1906 (Ben-Menahem *et al.*, 1976) and most recently in 1927 when the strongest earthquake of this century so far, with a magnitude of 6.2, struck the country, and at Damiya the banks of the river caved in, forming a barrier and causing the stretch of the Jordan down-river to run dry. It took several hours for the water 'heaped up' above the barrier to break through and renew the flow.

Avni *et al.* (2002) provide conflicting evidence suggesting the Jordan River did not dam as a consequence of the 1927 event, a re-evaluation of available data, including police reports, not supporting popular opinion in this regard. Nevertheless, occasional damming of the Jordan River as a result of landslides caused by seismic activity remains plausible over the historic past, as well as throughout antiquity. Psalm 114 summarises the situation admirably:

> The sea saw it, and fled: Jordan was driven back.
> The mountains skipped like rams, and the little hills like lambs.
> What ailest thee, O thou sea, that thou fleddest? Thou Jordan that was driven back?
> Ye mountains, that ye skipped like rams; and ye little hills, like lambs?
> Tremble, thou earth, at the presence of the Lord, at the presence of the God of Jacob;
> Which turned the rock into a standing water, the flint into a fountain of waters.

At Tell es-Sa'idiyah the excavations have disclosed the site to be of great antiquity with remains dating back to the Early Bronze Age of the third millennium. By 2900 BC the city was already a large, prosperous and flourishing community, with industries allied to the commercial production of olive oil and wine, and for the manufacture of textiles directed towards an export market. Fifteen hundred years later, during the New Kingdom period of Egypt (1600-1100 BC) Tell es-Sa'idiyah appears to have been a military centre for the Egyptian army, but this would have been several centuries after the destruction of Sodom and its satellite cities.

Early Trade

Some of the earliest trading links between Canaan and Egypt are described by Connan *et al.* (1992):

> The earliest material evidence for this trade is from the Chalcolithic period (4th millennium BC). A few pieces of Egyptian pottery were found in northern Sinai; a fragment of an Egyptian alabaster vessel is known from Ein Gedi, near the Dead Sea; and mollusc shells from the Nile river were found in southern Israel. (Gonen 1989)
>
> Rizkana and Seeher (1989) have reviewed the economic implications of the Maadi excavations in Egypt and shown that Chalcolithic strata contain imported items from Canaan. According to these authors, the items of Canaanite origin in Maadi include pottery, basalt bowls and spindle whorls, flint scrapers and blades, and perhaps copper and copper ore. The collapse of the Chalcolithic culture towards the end of the 4th millennium did not sever this trade connection. The booming trade between Egypt and Canaan in the Early Bronze period probably came to an end with the destruction of the Early Bronze cultural stage, and, although some trade may have continued, a major revival occurred only hundreds of years later.

One of the most informative descriptions of trade by Canaan, before its conquest by Joshua and the Israelites, is given by Kuhrt (1995) where she writes in connection with Ugarit:

> Given the rich documentation, both textual and archaeological, it is tempting to exaggerate Ugarit's commercial wealth and function. But it was only one of a number of similar states dotted throughout the Levant, along the coast and further inland; scholars simply happen to know a lot about it because of the wealth of written material, but Ugarit was not unique – it was one of many, though probably among the most important and prosperous. Its large archives show that the population was, on the evidence of personal names, made up of 'Canaanites' and Hurrians [*people of Northern Mesopotamia*]. The international links of the city-state are reflected by texts written in several foreign languages: Akkadian, the diplomatic language of the time, items inscribed with Egyptian hieroglyphs, some Hittite and Hurrian, and a small number of Cypro-Minoan texts. But most important (and interesting) is the existence and extensive use of a local alphabetic script – the earliest fully known one so far.

The script to which Kuhrt refers is what has been identified as Canaanite, a form of West Semitic. It used wedge-shaped signs (see Figure 26) impressed on clay tablets with a stylus, and though similar are not identical to other cuneiform

26. Ugarit script on clay tablet (Syrian Public Domain).

scripts, e.g. the Sumerian script illustrated on Figure 23. Because they were inscribed on clay, rather than more perishable materials, they have survived, and permit an insight into the daily life of the region. Kuhrt notes that apart from references to administrative, legal, and economic issues there is a substantial body of literary evidence suggesting close parallels to some of the poetry of the Old Testament.

The foregoing implies Ugarit to have been economically prosperous, with its wealth gained from an extensive trade throughout the eastern Mediterranean, extending to Cyprus, Egypt and Greece. However, it was not the only Canaanite city involved in such trade, coastal cities such as Tyre, Sidon, Byblos and Beirut, among others, were also centres. The notion that 'Canaan' and 'Phoenicia', if not synonymous, were certainly interlinked is thus underscored. The distant voyaging of the Phoenicians, and their settlements in North Africa, Europe and elsewhere during the first millennium, is testimony to their endurance as a stock. Some of this maritime expansion, whether in trade or settlement, may have been spurred by the Israelite conquest of Canaan following the Exodus.

The local availability of good quality timber placed the Phoenicians in the enviable position of becoming master ship-builders during a period when few nations sailed the seas with confidence. Timber was also in great demand by both Egypt and Assyria, and would have been one of the major exports of the Phoenicians, either as a bulk commodity or a value-added product, e.g. furniture. As the politics of the region changed, with the ascendancy of either Egypt or Assyria over the other, both generally hostile to each other, so the emphasis of trade conducted by Phoenician merchants would have fluctuated. In some instances, however, the Phoenicians acted as middlemen in direct Egyptian-Assyrian trade in commodities such as *frit* (a substance used in glass manufacture, and popular because of its blue pigment), but also in the supply of commodities such as alum, dyes, linen and, possibly, honey and wine (Kuhrt, 1995).

Kuhrt's comments regarding Old Testament poetry on the Ugarit tablets are of peculiar interest, as they tally with those of Shea (1979) in connection to the Ebla tablets. Shea also noted a textual similarity with some of the verses from the Old Testament. Specifically, two dozen verses could be identified with Genesis, a dozen each with Job and Psalms, and a half-dozen each with Isaiah, Proverbs and the Minor Prophets. This suggests the writers of the Old Testament had recourse to a wide extant literature.

The rise of an urban culture in neighbouring countries opened up trade routes across Canaan. Flanked by desert to the east and the Mediterranean on the west the land became a vital cross-road in trade with Egypt. Later it was destined to become a battlefield across which warring nations would roll back and forth. The land is still in contention today.

Subsistence agriculture began to change with new trading patterns. Specific crops and manufactured items began to be identified for export, and the resulting commodity market reflected a demand that was international as much as it was local. Wine, olive oil, grain, and salt are obvious commodities that would have commanded a good price in the urban centres of Egypt, but there was also bitumen, a material of considerable economic importance. As has been noted there is evidence of an extensive trade in this commodity, one that can be traced back to the Early Dynastic Period (c. 3000 BC) in Egypt, and which is the subject of the following chapters. Figure 27 shows the major trade routes across Canaan originating in the Dead Sea region.

Figure 4 (Chapter 1) indicates the oasis-like settlements fringing the southern part of the Dead Sea. These settlements, based upon a sustainable agriculture, would have been of vital importance to Sodom, whose prime activity was the harvesting and processing of bitumen, and to other centres in the region. The perception of Sodom and its satellite cities is that of a sprawling industrial complex centred around the South Basin of the Dead Sea. By the time of its demise Sodom could well have been at the 'heart' of Canaan by its huge contribution to Canaanite commerce.

As a consequence of trade, and the struggle for dominance that became inevitable, would have arisen the need for security. At Arad there is one of the best preserved sites of a settlement of the Early Bronze Age. Substantial stone walls with bastions indicate it to have been purposely designed and constructed to resist a large besieging force. Similar concerns appear to have dominated defensive measures elsewhere throughout Canaan as walls grew thicker in tandem with increased urbanisation. The indications are that this was not a restful period, and though internal strife may have led to much of the unrest, external forces, particularly Egyptian, may have become involved where trade links with Egypt were under threat.

The lack of contemporary written records in Canaan during the Early Bronze Age has led to the conclusion that the land was divided into regions under the suzerainty of major urban centres ruled by a 'king'. Increasing commercial prosperity of Canaan inevitably would have led to a concentration of political influence; a similar far-reaching influence of Ebla has been noted earlier. In this respect the 'cities of the plain' – Sodom, Gomorrah, Admah, Zeboiim, and Bela (or Zoar) – are likely to have represented a confederacy of relatively wealthy 'city states', though the term 'city' is probably too grandiose to reflect their actual populations. Canaan can therefore, be imagined as a vast complex of small 'kingdoms' in a state of constant flux as a result of changing trade patterns and fickle politics.

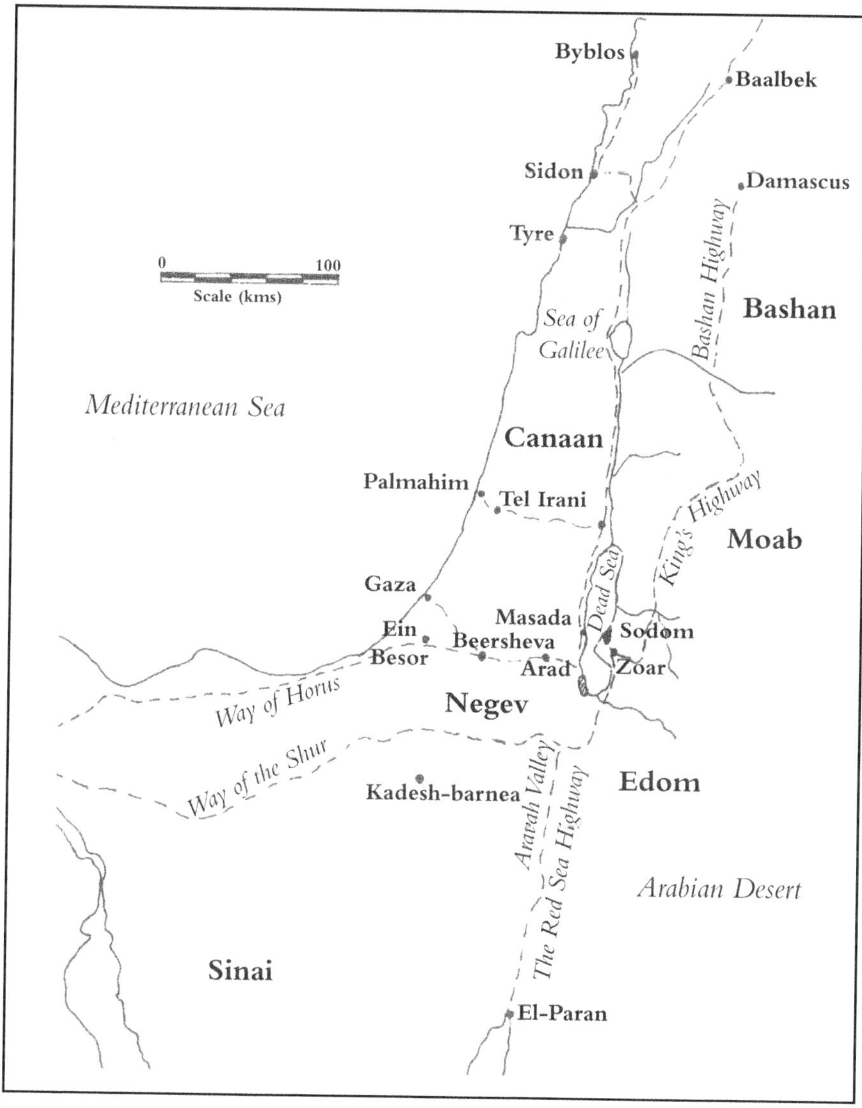

27. Map of Canaan and trade routes out of the Dead Sea.

Egypt was a major destination for the exports of Canaan whatever their nature. Though much trade was seaborne and followed the coast, a significant trade was also conducted along overland routes using donkey caravans. The preference for any route might well have reflected the perceived risks, as the conduct of long distance trade was a hazardous occupation at best. The perils besetting travellers then were likely no less than those of more recent times. On February 5, 1701 *The English Post*, London, reported the following:

> Letters from St John de Acre [*Akko*], a port town of Jerusalem, dated Sept 24, advise that 20,000 Arabians attackt and plundered the caravan in the desert going from Mecca to Damascus, consisting of 5000 mules richly laden, killing the Bassa who accompanied them, with his wives, and a great number of the travellers besides, most of the remainder perishing in the wilderness, the whole loss being computed at 70 tons of gold, but the caravan that went to Egypt escaped.

Early Bronze IV and Economic Recession

The Early Bronze Age has been subdivided to reflect changing phases of pottery manufacture considered indicative of cultural shift. Early Bronze I (or EBI) lasted 250 years from about 3100 to 2850 BC, followed by EBII and EB III, which occupied another 450 years until about 2400 BC. It was during EBII-EBIII that Canaanite culture appears to have fully flowered, and then at the end of EBIII it began to wilt. The following phase, known as EBIV, lasted a further 400 years, and has confronted scholars with a number of difficulties for the period was marked by considerable turmoil. So many changes occurred that the period has been associated with a major cultural discontinuity that overturned previously well-established patterns of trade, commerce, and settlement. Tubb (1998) writes:

> Characterization and understanding of the fourth phase, EBIV (2400-2000 BC), eluded scholars for nearly a century, and only comparatively recently has a consensus been reached which allows this enigmatic phase to be seen as a logical continuation of the culture of EBII-III. For so pronounced are the changes apparent in the material culture, settlement pattern and economy of EBIV that many scholars believed that they could be explained only in terms of a complete cultural break, resulting from widespread destructions of EBIII towns and involving mass invasions of nomadic barbarians. (Kenyon, 1979)

A great deal of controversy has surrounded the perceived cultural break at the end of EBIII to which Tubb refers, and he notes that much of the reasoning for the break has been now largely abandoned. More realistically, Tubb favours a reason for the cultural break as being based upon changes to the economy. Such a conclusion is supported by the implications of Neev and Emery (1995), who have written on this aspect of Canaanite history in the following terms:

> Evidence of a pattern of widespread destruction and abandonment is present at many Early Bronze III sites. From an archaeological point of view, this age witnessed one of the most sweeping disruptions in its history (Wright, 1961, p86; Kenyon, 1979, p134; Rast, 1987, pp194-95). This crisis at the close of

> the Early Bronze age was not limited to this country and adjacent regions but occurred as far away as the coasts of Syria and Lebanon. (Gophna, 1992) Transitions from First to Second Kingdoms in Egypt and from Sumerian to Akkadian rules in Mesopotamia are parallel examples. Perhaps a common causative factor such as physical disturbances induced social restlessness. Other authors consider that a cultural-demographic-economic process in Canaan could have been the only cause for the almost total cessation of maritime trade with Egypt during the end of the Old Kingdom. (Esse, 1989)

The disruption to trade between Canaan and Egypt as a consequence of the inexplicable destruction of Sodom would most certainly have provided a sound reason for such a cultural break.

Hiatus in Trade

It is a basic premise of this book that bitumen constituted the mainstay of the export trade of Canaan to Egypt, the harvesting of bitumen from the Dead Sea being the *raison d'être* for the existence of Sodom and her satellites, and one which contributed richly to the Canaanite economy. Following the destruction there would have been a devastating disruption of trade, leading to a total abandonment of the bitumen industry, and other allied industries. This would have resulted in the commercial disruption of Canaan, with inevitable consequences for the region as a whole, particularly the trade with Egypt. This could well have constituted a major cause leading to the rapid decline of Early Bronze culture in the region as noted by Connan *et al.* (1992) and Tubb (1998). The latter avers that Canaanite civilisation reached its zenith during the period c. 2800-2400 BC, following which there was an inexplicable mass emigration from the Dead Sea valley. The picture Tubb paints is suggestive of the region having suddenly lost its primary economic base, and this would have been certain with the collapse of the bitumen trade, which would have been inevitable following the loss of Sodom and its satellite cities. Tubb writes:

> Towards the end of the third millennium, the Old Kingdom in Egypt faltered and then collapsed, leading to a total cessation of trade links. The economic basis of Early Bronze Canaan, which had been dependent upon this commercial relationship, was therefore swept away. The effects, as might be expected, were profound. The loss of the Egyptian markets meant effectively that the Canaanite economy, which had been geared towards productive agriculture and trade, was no longer appropriate, and as a consequence urban settlement could no longer be sustained.

Once the importance of the bitumen trade with Egypt is realised a completely different perspective appears as to the cause of the economic collapse of the region, and the subsequent emigration out of Canaan. The trade in bitumen, and other associated raw commodities, can be perceived as being a major engine of growth to the Egyptian economy of the Old Kingdom. The cessation of that trade, due to the cataclysmic events which led to the loss of Sodom and her satellite cities, would have had devastating consequences. Rather than the economy of Canaan faltering due to political strife in Egypt, the reverse likely was more correct – Egypt was thrown into turmoil because of what had happened in Canaan. Once this argument is accepted, the reason becomes clearer for the abandonment of Canaanite centres during EBIV, and the outward migration of their inhabitants into Egypt and harsher regions of Transjordan and the Negev to adopt more pastoral livelihoods.

There is an opinion that Canaan had little to export in return for the many trade items of Egyptian manufacture encountered in archaeological excavations. Agricultural products such as wheat, barley, honey, wine and olive oil are obvious. Until lately the bitumen trade has been largely ignored, however, with modern analytical techniques it has been now possible to identify bitumen products in the excavations at Maadi, dating 3900-3500 BC, and proving them of Dead Sea origin. Though bitumen many have been the mainstay of industrial materials produced in the Dead Sea region, other commodities, such as natron and carnallite, could also have been exported and of equal importance to the Egyptian economy.

The disruption of this trade as a consequence of the disaster that befell Sodom and her satellites would have been unlikely to have affected any parallel trade between Northern Canaan and its neighbours, at least to any major extent. And this appears to have been the case, for the EBIV phase of ancient Ebla is indicative of a community that continued to flourish commercially and expand culturally.

The Early Bronze Age terminated turbulently in both Egypt and Southern Canaan, but this was followed by the eventual restoration of political and economic stability over a long period of time. During the ensuing Middle Bronze Age (about 2000-1750 BC) Egypt experienced a large influx of peoples, who seized control of parts of the Nile delta establishing political and military control which resulted in the creation of their own dynasty. The Hyksos, as they are now known, exercised some degree of control over the whole of Lower Egypt for a brief period, their influence being generally recognised as falling within the overall period of 1750-1550 BC. The Hyksos were later expelled by armies from Upper Egypt.

War is a means of establishing political control over trade and resources, and bitumen represented a desirable commodity, as much in antiquity as now. The events described in great detail in Chapter 14 of Genesis relating to the 'battle of the kings' in the Vale of Siddim, which have been discussed earlier in some detail, must be viewed with above average interest, especially the phrase 'twelve years they served Chedorlaomer, and in the thirteenth year they rebelled.' Over the millennia the land of Canaan has witnessed many invasions of hostile armies – including those from Egypt, Assyria, Mesopotamia, Persia, Greece and Rome among others, many of which had the objective of controlling bitumen production.

Religion

In matters of religion the Canaanites worshipped a vast array of gods and goddesses. Supreme of these was El, a remote god and father of all other gods. However, as far as everyday life was concerned it was Baal who was most influential, and who appears in various guises, together with his sister, Anat, the goddess of war and love. Also important was Astarte, goddess of fertility, and Lilith, goddess of wind. Figure 28 is a figurative illustration of Moloch, a god associated with child sacrifice through the exhortation of Leviticus (18:21) 'And thou shalt not let any of thy seed pass through the fire of Moloch'.

Parallels between the various divinities worshipped by the Canaanites can be observed elsewhere, most notably in Mesopotamia, Assyria and Egypt. Though Canaanite ritual undoubtedly involved some sacrificial ceremony, as evidenced by the worship of Moloch, in general human sacrifice does not appear to have been widespread. Though ossuaries have been discovered containing remains of children, some indicating cremation, it is uncertain whether these were a result of normal death, disease or sacrifice. Genesis (22:2-13) relates the command from God to Abraham to make sacrifice of his son, Isaac and the events which followed:

> Take now thy son, thine only son Isaac, whom thou lovest, and get thee into the land of Moriah; and offer him there for a burnt offering upon one of the mountains.... And they came to the place which God had told him of; and Abraham built an altar there, and laid the wood in order, and bound Isaac his son, and laid him on the altar upon the wood. And Abraham stretched forth his hand, and took the knife to slay his son.

Historically, this story precedes the conquest of Canaan by the Israelites following the Exodus. However, elsewhere in the Old Testament there are hints of human sacrifice, not by the Israelites, but by those adherents to the old religion of Baal. In Jeremiah (19:5) we read the words of God:

> They have built also the high places of Baal, to burn their sons with fire for burnt offerings unto Baal, which I commanded not, nor spake it, neither came it into my mind.

It is understandable that those who compiled the books of the Old Testament may have wished to differentiate between the monotheistic rites of the Israelites from those of the Canaanites, who still lived in the conquered land. As mentioned the Canaanites worshipped a multitude of gods, several dozen in fact. Many of these would have been considered pagan in comparison, if not barbaric. However, as in all pluralistic societies, there would have been temptations to worship new gods, especially if the old ones didn't perform as promised. This attitude is reflected by the following passage in Judges (2:11-13):

28. Illustration of the Canaanite deity Moloch from an 18th century German Bible.

> And the children of Israel did evil in the sight of the Lord, and served Baalim.[1] And they forsook the Lord God of their fathers, which brought them out of the land of Egypt, and followed other gods, of the gods of the people that were round about them, and provoked the Lord to anger. And they forsook the Lord, and served Baal and Ashtaroth.

The strife between the monotheism of the Israelites and the plurality of the Canaanites is exemplified by the story of Gideon, who was chosen by God to save Israel from the hand of the Midianites, and his cutting down of a grove sacred to the Canaanites one night as related in Judges (6:28-30):

> And when the men of the city arose early in the morning, behold the altar of Baal was cast down, and the grove was cut down that was by it. . . . And they said to one another, Who hath done this thing? And when they enquired and asked, they said, Gideon the son of Joash hath done this thing. Then the men of the city said unto Joash, Bring out thy son, that he may die: because he hath cast down the altar of Baal, and because he hath cut down the grove that was by it.

Furthermore, in Kings I (16:30-33) we read:

> And Ahab the son of Omri did evil in the sight of the Lord... And it came to pass . . . that he took to wife Jezebel the daughter of Ethbaal king of the Zidonians, and went and served Baal, and worshipped him. And he reared up an altar for Baal in the house of Baal, which he had built in Samaria. And Ahab made a grove; and Ahab did more to provoke the Lord God of Israel to anger than all the kings of Israel that were before him.

Again, in Kings I (19:18) God appears to Elijah and speaks as follows:

> Yet I have left me seven thousand in Israel, all the knees which have not bowed unto Baal, and every mouth which hath not kissed him.

---

1. Note: *Baalim* is interpreted as meaning images, or figurines, of Baal.

In Hosea (13:1-2) we read also:

> When Ephraim spoke trembling, he exalted himself in Israel; but when he offended in Baal, he died. And now they sin more and more, and have made them molten images of their silver, and idols according to their own understanding, all of it the work of the craftsmen: they say of them, Let the men that sacrifice kiss the calves.

It is obvious from these excerpts from the Old Testament that there was a struggle for supremacy between the two religions, that practised by the newcomers and that practised by the vanquished. Since the conquest of Canaan by Joshua can be dated to about 1450 BC following the Exodus from Egypt, and the Books of Jeremiah and Hosea can be dated to about 600 BC there was almost a thousand years of conflict between the two. As history shows, the Israelites prevailed in this conflict.

In Kings I (14:24) we read a verse possessing strange connotations. It reads:

> And there were also sodomites in the land: and they did according to all the abominations of the nations which the Lord cast out before the children of Israel.

The context is unclear as to whether this may be interpreted as referring to Sodomites, i.e. former residents of Sodom and its vicinity, or whether it refers to those practising sodomy. The text is more likely to relate to the latter since it is generally considered to have been written about 960 BC, which places it some 1400 years after the demise of Sodom itself

Textual evidence has also been found suggesting temple prostitutes were active amongst the priestly classes. There has been some suggestion that the local 'king' was also the 'high priest', and this would not seem unreasonable.

Temple construction and layout evolved with the progress of time, and though few remains of Early Bronze Age temples exist, those of the Middle Bronze Age indicate a similarity to temples of Egyptian design. It might be expected that temples in Sodom would have borne a similarity to those of an earlier age, possibly reflecting a Mesopotamian influence. One of the most famous temples ever to be unearthed was that of the ziggurat of Ur in southern Iraq by Leonard Woolley. Figure 29 depicts a diagrammatic reconstruction of the temple, based upon a sketch by Woolley (1939), showing how it might have looked at the time it was built by Ur-Nammu (2112-2095 BC), or his son, Shulgi (2094-2047 BC). Figure 30 is a recent photograph of the ziggurat. The facade to the lower portion has been rebuilt around the crumbling core, which may be seen protruding from the top. The main staircase (to the right) is much as it might have appeared when priests trod them to ascend the upper platforms.

The original ziggurat consisted of a core of mud bricks covered with a thick 'skin' of baked bricks set in bitumen. The base measured 60.5 by 43.0 metres.

29. A reconstruction of the Ziggurat of Ur based on a drawing by Leonard Woolley (1939) (US Public Domain)

Once it possessed three tiers, though only the first and part of the second tiers have survived. The height of the original structure is about twenty metres. Mitchell (1988) writes this about it:

> The surviving part . . . has roughly the form which it had in the time of Ur-Nammu, king of Ur 2112-2095 BC, but remains of later date show that in the time of the last Babylonian king, Nabonidus, 555-539 BC, it was rebuilt to a considerably greater height. Moreover there can be little doubt that Ur-Nammu's structure contains previous towers of decreasing size going back to a small temple, probably on a platform, of much earlier date, now buried in the core. This is clear from excavations at other sites such as Eridu, but it cannot be verified at Ur without destroying the ziggurat as it now survives.

It would be expected from this that there would have been temples at Sodom which were of similar construction. How magnificent they may have been in comparison we are probably never likely to know. As in all things change accompanies progress, and the temple architecture of the Late Bronze and Iron Ages reflects greater sophistication in the ability to construct grander edifices for worship. The temples at Sodom would be expected, as a consequence, to be less magnificent than later counterparts elsewhere.

Common Life

Life in Canaan amidst the humbler classes probably differed little from life in many parts of the region today. It can be expected to have been a harsh hand to mouth existence, with a short life expectancy for most. The priestly classes,

30. The reconstructed facade of the Ziggurat of Ur – the remains of the Neo-Babylonian structure protrudes at the top (US Public Domain).

however, would have had their abodes within the temple or its confines. As a consequence their living quarters would have been better constructed, and more lavishly adorned than those inhabited by the general populace. In contrast these would have been small shed-like cottages constructed from locally available materials which, in Canaan, would have been of uncarved stone, probably in the form of fieldstone and rubble. Naturally, no evidence remains of roof construction which, likewise, may be expected to have used local materials, probably a reed thatch supported by timber joists. Roofs of larger buildings, and those inhabited by the wealthier classes, may have also incorporated a coating of bitumen to render the roof watertight.

One domesticated animal, which would have relieved the burden of everyday toil for the lower classes, was the ass, or donkey. The African wild ass began to be domesticated some time after 4000 BC, and was used extensively in Egypt as a beast of burden, and still is in parts of the Middle East. Weighing up to about 250 kilograms the ass is a small animal, capable of carrying between 20 and 30 percent of its own weight, and surviving on low energy fibrous material, its stomach being peculiarly adapted to breaking down almost inedible vegetation. It was, therefore, ideally suited for survival in the marginal scrub lands and hilly terrain in proximity to the Dead Sea.

Tel Arad

The Canaanite city of Arad, which was founded about 3000 BC, was a major trading centre on the route from the Dead Sea to Egypt and the Mediterranean coast. This ancient city has been uncovered by archaeologists, and is controlled by the National Parks Authority of Israel. An information sign on entering Tel Arad reads as follows:[2]

2. See also Figure 31.

Tel-Arad

The area of the Canaanite city was 25 acres (about 100 dunam) and was encircled by a 1200 metre long wall with many towers, gates and posterns. The city was founded 5000 years ago and existed until about 2650 BC. In the city there is a network of streets through which rainwater flowed into a central reservoir. Piazzas and a series of public buildings including temples with courtyards in which an altar and a cult-basin were uncovered. A palace and many dwelling houses, where benches, cooking platforms, mortars and grinding stones were also found. The style of dwelling houses in Arad is uniform and known as the 'Arad House'. A model house was recently reconstructed in the south of the city.

The citizens were engaged in agriculture without irrigation, sheep and goat raising and trade with Egypt and Sinai. Animal statuettes, seals and many tools were found here. An Israelite fortress, built in the ninth to tenth century BC is situated in the north-east corner of the Canaanite city.

Bab edh-Dhra

One of the pioneering expeditions of the twentieth century, undertaken in the search for Sodom, was that led by Melvin Kyle in 1924. In his book *Explorations at Sodom* (1928), he tells of a Roman road leading towards a ford crossing to the western shore of the Lisan Strait. He writes:

> The Romans did not build roads to nowhere. That road led down to a bridge or a ford or a ferry, or simply to a continuation of the road itself right across to the western side of the valley. Arabs of Kerak have told Père Vincent that they still remember a ford at that place.

One of the expedition's findings, and perhaps the most exciting, was the discovery of the site of Bab edh-Dhra, and the realisation this might represent a Canaanite settlement of some note, the expedition concluding that it might represent that of Sodom itself. As this was an epic archaeological find the liberty will be taken to quote Kyle's observations at some length:

> Of the greatest interest, as well as for the light it sheds upon the culture of Sodom and Gomorrah, was the discovery . . . of a vast open-air settlement of the Early Bronze Age (third millennium BC) at Bab-ed-Dra [Bab edh-Dhra], five hundred feet above the Dead Sea on the road from Mezra'a to Kerak. The settlement contained a fortress over a thousand feet in length, surrounded by a massive wall and revetment.[3]

3. The former 10-15 feet thick, the latter some 15 feet high, were sufficiently well preserved.

31. The ruins of the Canaanite city of Tel Arad.

Just south of the fortress was this extensive open-air settlement consisting of hearths and enclosures, over which booths were probably erected. A few minutes walk to the east of the fortress, on the edge of the open-air settlement is a group of six sacred pillars, or masseboth, with the fragments of a seventh, all prostrate on the ground. The limestone monoliths must have been dragged for miles to be set up as masseboth.[4]

From the inside of the fortress and the open-air settlements to the south of it, we collected several thousand flint artefacts, besides several boxes full of pottery, mostly characteristic shards. The artefacts consist mainly of knives, some very fine, of scrapers for hides, of awls, and sickle edges, etc. The pottery is a very representative collection of types from the Early Bronze Age and the beginning of the Middle Bronze Age and covers several centuries, perhaps nearly a thousand years, from the middle of the third millennium to not later than the eighteenth century BC.

The pottery is mostly handmade, but is of good quality, and consists of several types: large plates or platters with an inverted rim, red in colour and highly polished or burnished; bowls with wavy ledge handles, small water decanters with vertical loop handles. Nearly all have flat bottoms.

The station of Bab edh-Dhra is thus of extreme importance for the comparative study of the ceramics, flint industry, and bronze culture of the second half of the third millennium.

---

4. Note: *Masseboth* are stones representing deities and, therefore, sacred and duly worshipped.

Commentary

This short summary of the history of Canaan, and insight into the life and religion of the Canaanites, is a poor substitute for more comprehensive texts available, e.g. Jonathan Tubb's excellent book *Canaanites: Peoples of the Past* (British Museum Press, London, 1998). However, the purpose of this chapter is to describe briefly the cultural backdrop of Canaan against which the destruction of Sodom, as depicted in Genesis, can be scientifically assessed.

If we accept the description in Genesis that the Plain of Jordan was 'well watered', and through archaeological and scientific research conclude the region capable, not only of producing olive oil, resins, fragrances and other tradeable commodities, but possessing a virtual monopoly on the supply of vital materials to Egypt, of which bitumen would be the most valuable, we must conclude also that the 'cities of the plain' were likely to be thriving communities by virtue of their association with Sodom. Some may have been more pleasant to live in than others, depending on the particular mainstay of commerce in which they engaged. If Sodom was the industrial centre for bitumen production, then it is likely to have been wealthier and more urbanised than most other centres in Canaan.

4. Bitumen in Antiquity – Sources, Recovery and Processing

Etymology and Nomenclature

There are numerous references to bitumen in ancient writings other than those contained in the Bible, where the most notable are contained in Genesis 6:14 (Noah's ark), Genesis 11:2 (Tower of Babel), Genesis 14 (Battle of the Kings), and Exodus 2:3 (Moses ark of bulrushes). However, translation difficulties apart, one of the great obstacles confronting us in our understanding of all ancient texts where reference is made to bitumen relates to the nomenclature used by the writers. This is perfectly understandable when we are in the fortunate position today of having mastered, at least to a major extent, the chemistry of petroleum and its derivatives, whether natural, refined, or synthetic in origin. Such enlightenment was denied to those early recorders.

Classical writers and many archaeologists, even in recent years, have used a variety of names to describe the black bitumastic materials, so widespread throughout ancient civilisations, which had numerous applications, and which were either of commercial application or great strategic significance. The words *asphaltos*, *bitumen*, *naptha*, *pitch*, *tar* and *slime* appear regularly in English translations with only passing reference as to whether the substance being described was solid or semi-solid.

The Hebrew Bible makes reference to substances such as *zephet*, *kofer* and *heimar*. Accordingly, there is some doubt as to the meaning of these terms, except for *heimar* which is taken to mean naturally occurring bitumen, or asphalt. In this context it is significant to note the location of natural bitumen deposits in the Dead Sea region, given on Figure 9 (Chapter 2), those located in the valley of Nahal Heimar possibly reflecting a philological root. When encountered in other ancient texts, such as those written in Sumerian, Akkadian, Egyptian, Arabic and Sanskrit, the difficulties become compounded. The Sumerians called it *esir*, the Akkadians *iddu*, and the Arabs had a variety of names – *sayali*, *zift*, or *qar*. All these terms must be considered as relating to crude, relatively unrefined, naturally occurring bitumastic materials.

The terms *bitumen* and *asphalt* are invariably treated as synonymous in today's world, the use of one over the other being reflective of linguistic preference. Etymologically it is of interest to note that though the term *bitumen* is considered to be derived from Latin, many consider it to have its origin in Sanskrit. The word *asphalt* on the other hand can be traced back to Greek. There is some consensus that both terms are possibly linked to *mumia* or mummy. A more precise definition of terms follows, but in the context of this book the term 'bitumen' is adopted throughout as it relates more properly to crude unrefined naturally occurring petroleum, whereas in modern parlance *asphalt* generally signifies a purer product, often refined or manufactured.

The earliest reference to petroleum as a specific substance first appeared in Agricola's *De Natura Fossilium* (1546) where he wrote:

> First of all the liquid . . . being similar to olive oil, is especially unctuous and has been named *oleum* (oil) by various writers at different times and is now called *petroleum* because it flows from rocks [*petra*]. This same black juice, when liquid is called *pix* (pitch) by others because of the similarity in colour to that of pitch. From this it is apparent that the name and nature of this substance was evident and well known to some and obscure to others. Thus many names coming from the vocabularies of different races have also been given to this same juice. The Babylonians call it *naptha*, the Samosatians *maltha*. The Moors, following the Arabs, as did several other races who were influenced by their learning, called it *hafral*. (Book IV)

Despite such problems in nomenclature, excavations of ancient sites, particularly those in Mesopotamia (modern Iraq) and Egypt, have furnished us with a wealth of information regarding the use of bitumen in those civilisations. It is regrettable, though understandable, that little attempt was made by early archaeologists to more closely identify the nature of the black substances they so frequently encountered in their excavations, i.e. the kind of bitumens used, their form and composition. However, this avenue of research has been pursued more recently by dedicated researchers with startling results, as spectrographic analysis of ancient bitumen products has now made it possible to link the bitumens concerned with their places of origin (Clapp, 1936; Forbes,1936). Though the term 'bitumen' has already been used in this text, to avoid any misconception it will be defined more exactly after Abraham (1929):

> A generic term applied to native substances of variable colour, hardness, and volatility; composed principally of saturated hydrocarbons, substantially free from oxygenated bodies; sometimes associated with mineral matter, the non-mineral constituents being fusible and largely soluble in carbon disulphide (CS_2).

It is obvious from this definition that the prime requisite of a 'bitumen' is that it is both a native substance and a saturated hydrocarbon regardless of its state as solid, liquid or gas. There is little evidence the ancient world made any extensive use of either liquid or gaseous hydrocarbons, either because such compounds evaporated quickly, or were too flammable. Exploitation of bitumen was confined, therefore, to mining into surface seepages, or digging of wells, often to great depths, to recover solid, semi-solid, or highly viscous substances that could be safely handled, treated and transported.

Bitumen Sources

The total loss of Sodom and its satellite cities has left us nothing in the form of ruins or other remains in the immediate proximity to the Dead Sea in order to examine the way bitumen may have been used in their construction, or to enrich their lives. This is not the case with Mesopotamia, where a rich legacy remains dating as far back as 3000 BC. A wealth of information from excavations at Ur, Babylon, and Eshnunna (Tell Asmor), in the form of mortars and mastics used in flooring, jointing, and water-proofing, has shown the engineers and artisans of those ancient times to have possessed great versatility in the use and application of bitumens and bitumen derivatives.

Outcrops of bitumen may be found scattered throughout the Near East, and are mentioned in historic writings. Though many were exploited in the ancient world, most have little economic significance today. Some of those mentioned are at Latakia, Beirut, the Yarmuk Valley, Hammath, Helwan (Egypt), Abu Doubar (Sinai) and Gebel Tanka (Suez). Strategically, the most important source in Mesopotamia encompassed the regions around Ramadi and Hît, on the southern bank of the River Euphrates. Here several hundred natural springs oozed bitumen. This was undoubtedly the centre of Mesopotamian production. It was also the source of the best bitumen, for it contained occluded water thereby rendering it pliable for packaging in a virtually 'ready-to-use' condition.

Since Id was the Akkadian name for Hît, and *iddu* was the Akkadian term for bitumen, the importance of Hît as a centre of production is evident. This is where King Tukulti Ninurta II (890-884 BC) wrote:

> In front of Hît by the bitumen springs, the place of the Usmeta stones, in which the gods speak, I spent the night.

The Usmeta stones were gypsum deposits impregnated with bitumen and sulphur from which gases were expelled through crevices, resulting in muffled noises which were interpreted as being the voices of the gods. (Bilkadi, 1984)

The inflammable nature of the gaseous and liquid components likely would have presented obvious perils to those engaged on its exploitation, but such risks would have been nullified by the commercial value of the solid product, a mainstay of trade. The Akkadians coined the term *naptu* (forerunner of *napht* or *naft* in Arabic) to describe the supposedly useless flammable crude oil from the highly prized *iddu*, or solid bitumen. Viruvius in his *De Architectura* (VIII:8-9) writes this in regard to various sources of bitumen in the general region:

> In Babylon, a lake of very great extent, called Lake Asphaltitis, has liquid asphalt swimming on its surface, with which asphalt Semiramis built the wall surrounding Babylon. At Jaffa in Syria and among the Nomads in Arabia, are lakes of enormous size that yield very large masses of asphalt, which are carried off by the inhabitants thereabouts.
>
> There is nothing marvellous in this, for quarries of hard asphalt are numerous there. So, when a quantity of water bursts its way through the asphaltic soil, it carries asphalt out with it, and after passing out of the ground, the water is separated and so rejects the asphalt from itself.

Though every kind of naturally occurring bitumen was to be found in the Hît-Ramadi region of Mesopotamia, most were of relatively high sulphur content (8-9 percent), and of variable purity (averaging 60 percent). This was in contrast to Dead Sea bitumen, which is about 98 percent pure with a somewhat lesser sulphur content.

Dead Sea Bitumen

Though bituminous limestones can be identified over a vast area of the Eastern Mediterranean extending south from Turkey to Palestine, and eastwards into Iran, none were as prolific, or renowned for the purity of its bitumen as those of the Dead Sea, which the Romans knew as *Lacus Asphaltitis.* Though Vitruvius mentions a lake of the same name, the lake to which he refers is located close to Babylon, and, therefore, cannot be the same.

Diodorus Siculus (*fl.* 60-30 BC) writes in his *Bibliotheca Historica* (XIX. 98) the following about the Dead Sea:

> It lies along the satrapy of Idumaea, extending in length about five hundred stadia [*92 kilometres*] and in width about sixty stadia [*11 kilometres*]. The water stinks and is exceedingly bitter so that fish cannot live in it, nor do any other aquatic creatures occur in it. Although large rivers of very fresh water flow in it, it remains bitter. Every year a large quantity of asphalt in pieces of more than 3 *plethra*[1] float in the middle of the sea, but often they are only 2 *plethra* in length. That is why the barbarians who live on the shore of the sea call the large pieces bull and the smaller pieces calf. When the asphalt floats in the middle of the water, it looks like an island to those standing on shore. The advent of the asphalt is heralded twenty days before its arrival, for all around the lake the stench is wafted by the wind over many stadia and all the silver, gold and copper in the neighbourhood becomes tarnished; but the tarnish disappears again when the asphalt rises to the surface. The district in the vicinity which is readily inflammable and which is pervaded by an unpleasant odour, makes the people's bodies ill and they die young.

1. A *plethron* is a measurement of length equal to a hundred Greek feet, or 40 paces, i.e. about 30 metres. Actual length might vary depending on the locality. The dimensions quoted by Siculus seem rather large.

32. A fragment of Dead Sea bitumen.

Strabo (*Geographica* XVI, 2.42) writes in similar terms, but there is a noticeable difference between his stated length of the Dead Sea compared to that of Diodorus Siculus. He writes:

> it extends parallel to the coast to a length of slightly more than two hundred stadia [*37 kilometres*], is deep to the very shore, and has water so very heavy that there is no use for divers, and any person walks into it and proceeds no farther than up to his navel is immediately raised afloat. It is full of asphalt. The asphalt is blown to the surface at irregular intervals from the midst of the deep, and with it rises bubbles, as though the water were boiling; and the surface of the lake, being convex, presents the appearance of a hill. With the asphalt there arises much soot . . . it tarnishes copper and silver and anything that glistens, even gold; and when their vessels are becoming tarnished the people who live round the lake know that the asphalt is beginning to rise; and they prepare to collect it by means of rafts made of reed.

Assuming both men are correct in the lengths they report, the implications are that Strabo's lesser figure was made when the Dead Sea was confined to the North Basin only, whereas that of Diodorus Siculus was made when the South

Basin had been inundated, if only temporally. Both men lived about the same time, Diodorus being probably the younger of the two. It is unknown whether either personally visited the Dead Sea, or whether they based their statements on third party reports.

The Jewish Roman historian, Josephus (AD 37-95), in *The Wars of the Jews* (IV:8) writes:

> The nature of the lake Asphaltitis is also worth describing. It is, as I have said already, bitter and unfruitful. It is so light [*or thick*] that it bears up the heaviest things that are thrown into it; nor is it easy for anyone to make things sink therein to the bottom, if he had a mind to do so. Accordingly, when Vespasian went to see it, he commanded that someone who could not swim have their hands tied behind them, and be thrown into the deep, when it so happened they all swam as if a wind had forced them upwards. Moreover, the change of the colour of this lake is wonderful, for it changes its appearance thrice every day, and as the rays of the sun fall differently upon it, the light is variously reflected. However, it casts up black clods of bitumen in many parts of it, these swim at the top of the water, and resemble in both shape and bigness headless bulls; and when the labourers that belong to the lake come to it, and catch hold of it as it hangs together, they draw it into their ships; but when the ship is full, it is not easy to cut off the rest, for it is so tenacious as to make the ship hang upon its clods till they set it loose with the menstrual blood of women and with urine to which alone it yields. This bitumen is not only useful for the caulking of ships, but for the cure of men's bodies, accordingly it is mixed in a great many medicines. The length of this lake is five hundred and eighty furlongs, where it is extended as far as Zoar in Arabia, and its breadth is a hundred and fifty.

The length of the lake as given by Josephus thus agrees well with that reported by Diodorus, and both estimates are greater than that of Strabo. Vitruvius, Diodorus, Strabo and Josephus were contemporary writers and all wrote about the turn of the millennium. To what extent they communicated and shared their observations is unknown, but their descriptions relating to the harvesting of the bitumen are consistent and bear the mark of respectability.

Josephus's observations regarding the application of urine in 'cutting' the slugs of bitumen when harvesting is of above average interest, as urine had numerous uses until recent times. Urine becomes mildly caustic on ageing, and was used extensively in cleaning wool and tanning as it possesses the property of dissolving greases; it was also used by housemaids for cleaning purposes; and on ships to combat lice. As will be seen later caustics also have a use in bitumen refining.

The comments of Strabo, where he writes 'it tarnishes copper and silver and anything that glistens, even gold' is also of interest, as pure gold does not tarnish.

Its tarnishing would suggest the gold concerned included a minor amount of copper. Pure gold is very soft, and it is a common practice to add small amounts of copper to render it more resistant to general wear and tear.

The unpleasant odours referred to by Strabo are, undoubtedly, compounds of sulphur or ammonia which are associated with most hydrocarbon deposits to greater or lesser extent. Uren (1946) writes:

> The petroleum odour of hydrocarbon gases associated with oil deposits is often masked by the stronger odour of hydrogen sulphide and sulphur dioxide with which they are contaminated in nature. A characteristic odour of ammonia is also often detected. The ammonia may be derived from naturally occurring ammonium compounds, such as the sulphate, by contact with hydrocarbons in either the liquid or gaseous phases. Interaction of the two sulphur gases with water sometimes results in the formation of native sulphur as a sublimate. Sulphuric acid similarly formed may give groundwater a slight acidity. In many gases the sulphur compounds are derived from sulphates by the reducing action of hydrocarbons. Anaerobic bacteria associated with groundwater in contact with petroleum has been shown to have the power to convert sulphates into sulphides.

Less credible than the aforementioned writers are the descriptions contained in *The Voiage and Travaile of Sir John Maundeville, Knight*, a book which has been proved a delightful forgery of the fourteenth century, for the traveller apparently never left the comforts of his library. 'Sir John' writes:

> And two myle from Jerico is flom Jordan and you shall wete the Dead Sea dapareth the land of Juda and of Araby and the water is right bitter and this water casteth out a thing called asphaltum, as great pieces as a horse.

Numerous republications of Maundeville's alleged travels have been made over the years, e.g. Maundeville (1836). The writings of Diodorus Siculus (and of 'Sir John' Maundeville) indicate bitumen rising to the surface 'in pieces of more than three *plethra*' (or 'in great pieces as a horse'). Indeed, the recent risings of 1955 fit these descriptions well, and in the 1930s reports suggested the bitumen attained weights of between twenty and thirty tons. Analysis showed the bitumen to fit all the characteristics of asphaltites bitumen. It may be concluded, therefore, that the bitumen described by Diodorus Siculus belongs to this class. On the other hand Strabo omits describing Dead Sea bitumen in terms of size, which would be a serious omission if he had indeed observed first-hand large pieces floating upon the surface. Instead he mentions the fact the bitumen is collected by means of rafts made of reed. This suggests that Strabo is describing asphaltic bitumen, or petroleum crude floating upon the surface. This implies two types of bitumen may have been harvested from the Dead Sea, but whether they had the same source remains uncertain.

Diodorus Siculus and Strabo refer to the 'stench', the 'tarnishing' of metals, and the short expectancy of life of workers engaged on harvesting the bitumen, whether it be in the form of large lumps, or viscous liquids or semi-solids. Undoubtedly the 'stench' is hydrogen sulphide (H_2S), an odiferous and poisonous gas associated with high sulphur crude or 'sour' oils, but carbon dioxide (CO_2), equally poisonous in large amounts but odourless, is also likely to have been emitted.

The harvesting of large floating pieces of asphaltites bitumen presents few problems, though the sizes reported, i.e. three *plethra*, about 90 metres in length, seems rather exaggerated. However, even weighing several tons they could have been towed into a suitable dock, where they would have been chiselled, or broken up, into smaller, more easily handled fragments. As the asphaltites is pure, and of relatively high fusibility (130°C), all that is required preparatory to transportation, is to remelt it to form suitably shaped briquettes, and to repack.

Nissenbaum (1978) describes the various types of bitumen encountered in the Dead Sea region. He says this of the floating blocks of asphaltites:

> The blocks are of various sizes, some weighing more than 100 tons. They may float on the lake surface or, more commonly, are cast up on the shore. The asphalt is very pure, has an ash content of less than 0.1 percent, and is characterized by high sulphur content (about 10 percent). The blocks are pitch black, with a very high lustre on fresh surfaces, and commonly have vesicular structures [*small bubble holes*] as well as broad surface striations. The specific gravity of the asphalt is 1.115, and it floats easily in the Dead Sea, whose waters today [*1978*] have a density of 1.230. The exact source of this asphalt is unknown. Surface exposures of asphalts [*bitumens*] are too small, or too impure, to provide the observed quantities of floating blocks so, presumably, the asphalt originates from the bottom sediment of the lake.

The fragment of asphaltites bitumen shown in Figure 32 was given to the author by Arie Nissenbaum. The vitreous nature of the substance is obvious, and from a distance the fragment has all the appearance of a shard of black glass. However, in addition to asphaltites bitumen a wide range of bitumen-based relics may be found along the shore from time to time, and Figure 33 illustrates the variations in some of the hand specimens collected.

The recovery of viscous liquid or semi-solid asphaltic bitumen floating upon the surface of the water requires greater ingenuity to recover, and then render it suitable for further processing or transportation. Tacitus (*Historiae* V.6) writes this about the harvesting of the bitumen:

> Those who make it their business to collect it, draw one end [*of the floating bitumen*] into their boats; the rest of the mass follows without toil or difficulty and continues loading the vessel, till the viscous substance is cut in two. . . . This extraordinary substance, floating in heaps up and down the

33. Pieces of natural bitumen collected from the shoreline of the Dead Sea (Courtesy Daniel Tzvi – Public Domain).

> lake, is driven towards the shore, or easily drawn by the hand; and when the vapour that exhales from the land, or the heat of the sun, has sufficiently dried and hardened it, it is cut asunder by wedges or the stroke of a hatchet.

Tacitus mentions the drawing of the asphalt into boats, and the fact that it is 'viscous' and is hardened by the 'heat of the sun'. Strabo refers to its collection using rafts made of reeds. Pliny in *Historiae Naturalis*, and Agricola in *De Re Metallica* (1556), also make reference to the harvesting of bitumen floating upon waters in Sicily and Germany in the following terms:

> It also occurs as a rich oily liquid in Sicily where it contaminates the waters of a spring at Agrigentium. The local folk collect it by means of bunches of reeds, to which it adheres very readily. (Pliny)

> Liquid bitumen sometimes floats in large quantities on the surface of wells, brooks and rivers and is collected with buckets or pots. Small quantities are collected by means of feathers, linen towels and the like. The bitumen easily adheres to these objects. (Agricola)

It would seem apparent from these accounts that two types of bitumen were harvested, i.e. asphaltites and asphaltic bitumen. The former would have been the hard blocks to which Nissenbaum refers, and which were of high purity. The latter would have possessed the consistency of a viscous semi-solid, which may be imagined as a black, gooey, slippery mass, the slugs of which would have congealed as time elapsed to form increasingly larger pods. The general description of this being *slime*, in some translations of the Old Testament, may, therefore, be peculiarly appropriate.

The two types of bitumen therefore contrast with each other in regard to size and shape. The asphaltites bitumen occurs in the form of large floating blocks of several tons weight, and these must be the 'islands' of Diodorus Siculus, or the 'great pieces of 'Sir John' Maundeville. They are obvious and unmistakable to the onlooker. Not so obvious is the asphaltic bitumen requiring greater diligence in its harvesting. Its consistency can be expected to have ranged from liquid, to a highly viscous semi-solid, to that of a hard platelet. As evaporation of the more volatile components proceeds with the elapse of time, so the material hardens. The method of harvesting would have been dependent to a large extent upon the stage reached in the natural evaporation process, the harder it becomes the more manageable it is to harvest. Though the use of nets in such a context is not mentioned by any of the ancient geographers, their use must not be discounted. The term 'fishery', often encountered in many writings connected with the harvesting of Dead Sea bitumen, may be an apt description.

Since 2003 a number of 'asphalt volcanoes' have been discovered on the sea floor in the Gulf of Mexico and off the coast of California, often in association with salt diapirs, i.e. salt formations thrusting upwards into overlying strata. The 'volcanoes', or seamounts, are essentially bitumastic in composition, the nature of the bitumen varying widely. Nissenbaum (1978) made the point that as the origin of Dead Sea bitumen is uncertain, it may be released from the bottom of the lake as a consequence of diapiric growth, He states that the temperature at which the asphaltites softens, and begins to attain fluidity, lies between 60° and 80°C, and the geothermal heat at the bottom of the Dead Sea is sufficient for its mobilisation. Strabo described the bitumen as 'a clod of earth which is liquefied by heat'. Nissenbaum writes further:

> The ancient writers also mentioned two attributes of the asphalt [*bitumen*] which do not agree with the modern examples. The first is the association of asphalt appearances with large-scale exudation of hydrogen sulfide. The second is the consistency of the asphalt. The asphalt known from the 19th and 20th centuries is a solid, unmalleable substance, whereas the ancient descriptions are of semifluid, viscous substances which could be solidified by using vinegar or by exposure to air.

The overall evidence concludes the existence of two types of bitumen. The first being the harder, purer, asphaltites rising from great depth in the waters of the Dead Sea, and having its source in the older Jurassic rocks; the second, being the viscous asphaltic bitumen seeping from the shallower overlying Cretaceous strata, which are known to bear hydrocarbons of a liquid and gaseous nature. Sporadic occurrences of impure bitumen may be noted impregnating rock outcrops in various parts of the region. It would not be unusual for changes in the frequency of the risings, and nature of their consistencies, to have changed over the centuries. Certainly, the rising of asphaltites bitumen in the Dead Sea appears to have become less frequent now than may be inferred from ancient writings. Peters *et al.* (2005) explain the intermittent nature of such discharges in the following terms:

> As petroleum is expelled, the pressure drops and fractures close temporarily. Silica or calcite commonly precipitate along closed fractures, helping to repressurize the system until the next episodic release of petroleum. This process continues until the fluid-generation rate diminishes below the point where fluid pressures are sufficient for expulsion.

The overall time period of this sporadic type of expulsion is, of course, dependent on such factors as the nature of the source, its depth, volume, the driving pressure involved, and the likely influence of seismicity. It is possible that the rise and fall of lake level, resulting in changes to hydrostatic pressure at sea bottom, may influence the expulsion of the bitumen from its source rocks. If this is so, the present trend in the fall of Dead Sea water level may result in the recommencement of periodic risings of bitumen 'bulls'. Figure 33 illustrates various pieces of bitumen recovered along the shore of the Dead Sea.

Boats and Bitumen

Bilkadi (1984) claims that the marsh dwellers, or 'Ubaids of Southern Mesopotamia, were the first to use bitumen for waterproofing their huts and boats, both being manufactured from reeds. In the construction of their huts they would bundle the reeds together with bulrush fibres to form columns, and weave mats to provide a roof. Interior walls might be plastered with mud, but bitumen would be used to render waterproof the external walls and roofs.

By 4000 BC the 'Ubaids were roaming the marshes of the Shatt-el-Arab in their rafts and boats of plaited reed covered with stretched animal hide and payed with bitumen, the earliest known remains of a reed boat, found in Kuwait, being dated to about 5000 BC. The use of reeds in the construction of floating vessels thus stretches far back into antiquity. A popular mode of water transport, prevalent in Southern Mesopotamia, was the *kuffah* or *guffah*, a large basket made from closely woven willow branches. Both surfaces, inner and outer, were often coated with bitumen for the safe transportation of goods, particularly livestock. This required a suitable rendering to the inside of the vessel, and for sea-going craft it was essential this be waterproof. The *kuffah* thus resembled a huge bowl, and could be built large enough, and sturdy enough, to carry up to three horses and several men. These were similar in construction to the reed boats constructed from the local *burdi*, a form of papyrus, though this is reputed to have been inferior to that growing in Egypt.

There is archaeological evidence (Bibby, 1962) of significant maritime trade between Mesopotamia and the legendary city of Dilmoun, believed to be Bahrain, stretching back over five millennia. The trade would have relied upon large ocean-going vessels, known as *ma-gur*. These were constructed of reeds. Today, there are many examples of reed rafts and fishing boats in use in the Persian Gulf, Bahrein being a notable centre, as well as farther afield in Africa and South America.

Southern Mesopotamia was devoid of timber from which to build boats, which explains the development of the *kuffah*. But this was not the case with the Dead Sea, for, according to Reifenberg (1955), deforestation of the region did not occur until after the Arab conquest (c. AD 700). However, there is no reason why the Sodomites would not have developed similar such rafts for their own peculiar applications.

Suitable timber to build boats for the bitumen fishing industry of the Sodomites, if not immediately to hand, at least was readily available. These boats would have to have been sturdily built, for they were required to do hard work. Therefore, it is doubted whether the rafts of reeds, as described by Strabo, were the only flotation methods of harvesting the bitumen. It is conceivable that boats constructed of wood, and rafts made from reeds, were used in tandem depending on the type of bitumen harvested. When blocks of asphaltites were being harvested wooden boats would have held the advantage, but when highly viscous asphaltic bitumen was being collected reed rafts would likely have been used, either alone or towed by wooden boats.

Pliny and Agricola both emphasise the use of feathers, linen, reeds and such materials to absorb bituminous liquids from water surfaces. The rafts of reeds described by Strabo would have served a similar purpose, their main application being to act as 'sponges' to mop up the liquid and semi-liquid bitumens floating on the surface. It must be doubted whether the Sodomites had any technical understanding of the physics of such phenomena as capillarity or surface tension, but through experiment they could have arrived at a simple means of harvesting this valuable resource.

Keller (1956) notes cuneiform texts have this to say about King Sargon, founder of the Semitic dynasty of Akkad in 2360 BC:

> I am Sargon, the powerful king of Akkad. My mother was a temple prostitute: I did not know my father. My mother conceived me in secret. She put me in a little box of reeds, sealing its lid with pitch. She put me in the river.

This is virtually identical to the story of Moses (Exodus 2:3):

> And when she could no longer hide him, she took for him an ark of bulrushes, and daubed it with slime and pitch, and put the child therein.

Processing of Bitumen

Because of their high purity, the large floating blocks of asphaltites would have required little processing once brought ashore apart from, perhaps, melting and recasting to form suitable shape and size for transport. The asphaltic bitumen, on the other hand, would have presented a greater challenge, for it would have been brought to shore in the form of impregnated reed rafts, or slithering masses of varying viscosity in the bottoms of boats. Processing would have necessitated recovery of the bitumen from the reed rafts, the removal of adulterating water, sand and grit, and, if there was an eye to producing a high quality end product, the removal of contaminants such as sulphur. This suggests the likely existence of a centralised refinery, primitive though such a refinery might be in comparison to those we know today. The 'slime pits' described in Genesis 14, by implication, suggests the location of such a refinery as being within the Vale of Siddim.

The technology used by the Sodomites in refining bitumen recovered from the Dead Sea is, of course, highly speculative. Agricola in *De Re Metallica* (1556) makes passing reference to the refining of bitumen, but it appears from the context he had no direct experience as he quotes Pliny, saying:

> The water, bituminous as well as salty, at Babylon, as Pliny writes, was taken from the wells to the salt works and heated by the great heat of the sun, and condensed partly into liquid bitumen and partly into salt. The bitumen being lighter, floats to the top, while the salt being heavier, sinks to the bottom. (Book XII)

The methods described by Agricola, following Pliny, may indeed have been practised. However, it is of value to give thought to other possibilities that may have been adopted by those early forerunners of the 'petroleum engineer', for there is little doubt the Sodomites, over the ages, had acquired considerable knowledge in the practice of their industrial art.

(1) The removal of highly viscous or solid bitumen impregnating vegetation may be facilitated by raising the temperature to make the bitumen flow more readily. This requires heating in a retort in the absence of air, in order to avoid ignition. The temperature to which the mass would have to be raised would be highly dependent upon the specific properties of the batch being treated, which could vary widely. Also stringent control would be necessary to avoid a conflagration, and that would require great experience on the part of the operator, or overseer.

For asphaltic bitumens it is likely the temperature for a sustained flow of melted bitumen would be no greater than about 60 to 70°C. This is in contrast with asphaltites from the floating blocks, which possesses a high purity of about 98 percent, and which melts at a significantly higher temperature of about 130°C.

(2) The removal of all adulterating materials to obtain a high quality product would have necessitated stripping the bitumen of sand, mineral salts picked up from the saline waters, excess water and all sulphur. The extent to which this was done is, of course, not known, but it is certainly feasible. The magnitude of success of sequential operations could have depended on the availability of caustic soda

or sodium hydroxide (NaOH), or sodium carbonate (Na_2CO_3). Alternatively aged urine may have been used as this is caustic. The possible manufacture by the Sodomites of caustic soda and other chemical products will be discussed in Chapter 6.

In principle the stripping operations may have been a precursor to what is now known as 'the cold water extraction process' for refining bitumen, which is of wide application, particularly in the treatment of tar sands. This involves the creation of alkaline slurry in aqueous solution containing small quantities of sodium carbonate, approximately 1.5 kilograms to every cubic metre of bitumen slurry. The alkalinity helps to break the surface bond of bitumen to sand, enabling the sand to be more quickly precipitated. The bitumen rises to the surface in the resultant froth. The addition of caustic soda also assists in the removal of sulphur compounds.

Considerable agitation is needed in the early stages of this process for stripping to be successful. Thus, a series of pits can be envisaged, the last pit in the chain being the one where the bitumen is raked off for air-drying.

(3) The drying yard would represent the final stage in the refining process. Here the bitumen is allowed to solidify before being cut into sizes convenient for transport or, alternatively, cast into moulds and then air-dried. If a pit was used it would be large and shallow.

The refining operations would have been very labour intensive. Working conditions would have been unhealthy, and the life-expectancy of workers is likely to have been short. Therefore, it is probable that slave, or impressed, labour, constituted the backbone of the workforce of the Sodomites.

Commentary on Recent Developments

Modern thinking regarding exploitation of bitumen resources in the Dead Sea region was summarised by Clapp (1936) when he wrote:

> [V]ast areas exist in which petroleum source rocks may be present and in some places they contain 10-16% crude oil. Certain strata are oil-bearing and liquid petroleum may have accumulated in some places. The area around Jebel Usdum [*Mount Sedom*] and el-Lisan [*the Lisan Peninsula*] has greater oil potentialities than the higher plateaus. . . . Whether commercial supplies exist in the Dead Sea area or not, the region is on record as having produced bitumens or petroleum since earliest known habitation and hence is of definite scientific and historic importance.

We now know that although there are hydrocarbon deposits in the Dead Sea region, at the time of writing none have been proven commercially viable, though it is likely that 'fracking' as commonly known will be useful in the future extraction of shale gas. It is humbling to think the bitumen deposits of the Dead Sea, exploited by the Sodomites over four thousand years ago, probably represent one of the first oilfields ever worked by man, but today is too small to be worked profitably assuming it has been properly delineated and its reserves quantified. Clapp's comment regarding the area around the Lisan Peninsula offering potential is of more than passing interest, as this is where the postulated Vale of Siddim and the bitumen refining process of the Sodomites would have been located.

There are some notable differences in the geology between the two shores of the Dead Sea which, likely, impact the ability of the rock strata to provide suitable reservoir conditions for the containment of hydrocarbons. The western (Israeli) side is dominated by low porosity limestones and shales of Cretaceous age. In such strata hydrocarbons are mostly confined to fractures, cavities and other defects within the host rocks, as the transmissivity of limestone/shale is low. On the other hand the eastern (Jordanian) side is dominated by older Nubian sandstones of the Paleozoic-Mesozoic, which possess porosities more favourable to the accumulation of liquid and gaseous hydrocarbons. The hydrocarbons on the Israeli side are in the form of sporadic occurrences of bitumen impregnated sediments, or as oil-shales, which require high capitalisations to process commercially though, as stated, 'fracking' may prove applicable. The Jordanian shore, on the other hand, is more attractive commercially, with hydrocarbons having proved recoverable from relatively shallow depth, however, it has the disadvantage of a complex geology which limits recoverable volumes contained in individual reservoirs.

The eastern shore of the Dead Sea has been foremost among the targets of petroleum exploration engineers over the past century, and a number of wells have been sunk with a view to assessing commercial viability. No major discoveries have been reported though it has been concluded hydrocarbon reserves exist at relatively shallow depth, but because of the high degree of geological structure the trap conditions are complex and individual reservoirs limited in volume. At present none appear to justify commercial exploitation. One well, the Ain El Hummar-2, indicated the presence of three types of crude oil of medium gravity (23.3 to 31.1° API), and medium sulphur content (2.14 to 2.65%) possessing isotopic and biomarkers reflecting similarity of source. The source was concluded as Triassic in age.

Though somewhat debatable, the source of the bitumen risings must be perceived as originating within the fractures associated with the various fault systems beneath the North Basin of the Dead Sea. It is possible that the degree of mobilisation, and periodic outpouring of the bitumen, may be a reflection of tectonic activity. In this it may have some parallel to Trinidad Lake asphalt, a highly productive deposit from which 150,000 tons are won annually.

5. Bitumen in Antiquity – Applications and Trade

Early Applications

The first practical use of bitumen probably originated with nomadic hunters realising they could better adhere flint arrowheads to the shafts of their arrows, as Connan (1999) reports the unearthing in the Syrian desert of flint fragments coated with bitumen, dating to 40,000 BC. Later, the adherent properties of bitumen would be used in the manufacture of sickles and knives, and for less utilitarian purposes in the decorative arts. It was to be used extensively for waterproofing containers, such as earthenware pots, storage jars, baskets, as well as for lining baths, cisterns, sealing of sarcophagi, and general caulking to pipes and gutters as well as waterproofing roofs of reed or palm. Numerous household objects, in the form of vases and jewellery have been found, in which shells, mother-of-pearl, and colourful stones are attached with bitumen.

For over three thousand years the centre of Mesopotamian bitumen production was at Hît on the Euphrates River, for a long time the head of navigation originating in the Persian Gulf. One of the first applications of bitumen, and one which would be of immense importance to the politics of the region, would have been in the water-proofing of boats, especially those constructed of reeds, for quality timber was scarce. This enabled the swampy marshlands of the south to support a larger population than otherwise, and facilitated the expansion of regional maritime trade.

There are three references to bitumen in the Bible which are of above average interest. These are:

(1) Genesis 6:14, when God instructs Noah to build his ark – 'Make thee an ark of gopher wood; rooms shalt thou make in the ark, and shalt pitch it within and without with pitch'.

(2) Genesis 11: 3-4, when the inhabitants of Babel decide to build a tower 'whose top may reach to heaven' – 'Go to, let us make brick, and burn them thoroughly. And they had brick for stone, and slime had they for mortar'.

(3) Exodus 2:3, when the mother of Moses wove a basket of bulrushes – 'And when she could no longer hide him, she took for him an ark

> of bulrushes, and daubed it with slime and with pitch, and put the child therein; and she laid it in the flags by the river's brink'.

These Biblical references to *pitch* and *slime* are generally considered synonymous, and in the terminology of this book both are considered as *bitumen*. Though Noah used bitumen as a caulking/waterproofing material, and the builders of the Tower of Babel used it as a jointing mastic, as we would use cement mortar today, its use in the basket of bulrushes is more fascinating. Bitumen, despite its acknowledged usefulness, is a black, sticky semi-solid to which the average woman is unlikely to have ready access, unless her husband or members of her family were in the building trade, where the substance was widely used. Therefore, it may be concluded the father of Moses, 'a man of the house of Levi', was probably a builder.

Natural occurrences of bitumen and associated hydrocarbons are widespread throughout the Middle East. However, it is now possible, by means of advances in analytical techniques, to trace the origins of bitumen products back to their sources when encountered on archaeological sites. A detailed account of the bitumen trade in antiquity, as revealed by molecular analysis of the constituents, has been given by Connan (1999). He writes:

> Ancient people from northern Iraq, south-west Iran and the Dead Sea area extensively used this ubiquitous natural resource until the Neolithic period (7000-6000 BC). Evidence of earlier use has been recently documented in the Syrian desert (Boëda *et al.* 1998) near El Kown, where bitumen-coated flint implements, dated to 40,000 BC (Mousterian period), have been unearthed. This discovery at least proves that bitumen was used by Neanderthal populations as hafting material to fix handles to their flint tools. Numerous testimonies, proving the importance of this petroleum-based material in Ancient civilizations, were brought to light by the excavations conducted in the Near East as of the beginning of the century.

Connan and Deschene (1995) mention the wide-ranging applications of bitumen in construction during Antiquity, making particular reference to the Darius Palace at Susa, the Hanging Gardens and Tower of Babel at Babylon, and other temples and ziggurats throughout Mesopotamia and Elam. One of the more interesting uses was for road surfacing, the processional avenue of Babylon being especially noteworthy. They also note the finding of lumps of bituminous materials as far away as Ra's al-Junayz in Oman, suggesting a wide-flung trade throughout the Persian Gulf. They also note the application of bitumen in the decorative arts, where it was used as:

> a widespread adhesive in antiquity and served to repair broken ceramics, fix eyes and horns on statues (e.g. at Tell al-Ubaid around 2500 BC).

> Beautiful decorations with stones, shells, mother of pearl, on palm trees, cups, ostrich eggs, musical instruments (e.g. the Queen's lyre) and other items, such as rings, jewellery and games have been excavated from the Royal tombs in Ur.

By any standards Babylon, during the reign of Nebuchadnezzar II (605-562 BC), was a magnificent city, for he was a prodigious builder and took great pride in its construction. Sprawled along the palm-fringed banks of the Euphrates, Babylon was to become a great metropolis renowned for its wide avenues, tall buildings, numerous temples and mighty walls, but all these were puny in comparison to the Tower of Babel that dwarfed everything, and that other wonder of the ancient world, the Hanging Gardens of Babylon. Hundreds of thousand of clay bricks bore his name and no city of the time surpassed its grandeur, neither Memphis, Thebes, nor Ur.

In its polytheism Babylon presented a bewildering array of gods for, altogether, there were fifty-three temples dedicated to the chief gods, fifty-five chapels to Marduk (god of sky and war), three hundred chapels to the earthly deities, six hundred for the heavenly deities, one hundred and eighty altars to Ishtar (goddess of love), one hundred and eighty for Nergal (god of the sun and the netherworld) and Adad (the storm god), and twelve altars for lesser gods.

Bitumen was used extensively throughout Babylon and its environs. It was used in paving, in the construction of bridges across the Euphrates, in dams, and for waterproofing drainage systems, channels and cisterns. Its ubiquitous use was a tangible exhibition of Babylonian dominance in the arts of construction, for it was Nebuchadnezzar who completed the construction of the Tower of Babel, which must count among the foremost structures of all Antiquity. Herodotus (c. 484-c. 425 BC) described it in the following terms:

> In the midst of the temple a solid tower was constructed, one stadium in length and one stadium in width. Upon this tower stood another, and again upon this another, and so on, making eight towers in all, one upon another. All eight towers can be climbed by means of a spiral staircase, which runs around the outside. About halfway up there are seats where those who make the ascent can sit and rest. In the topmost tower there is a great temple, and in the temple is a great bed richly appointed, and beside it a golden table. No idol stands there. No one spends the night there save a woman of that country, designated by the god himself. The priests told me that the god descended some times to the temple and joined her . . . but I cannot believe this.

The Tower of Babel was, therefore, a massive stepped structure, one that towered ninety metres above the rooftops of Babylon, each of its facades a different colour, its bricks cemented with bitumen. Its great height may have enabled the priests to better communicate with the deities that dwelt in Heaven above, but it was only one of many ziggurats that dotted the plains of Mesopotamia though, unarguably, it was the most colossal requiring millions of bricks and many hundreds of tons of bitumen in its construction.

Bilkadi (1984) reproduces Nebuchadnezzar's own eloquent description of the building of his palace:

> In Babel, my favourite city that I love, was the palace, the house, the marvel of mankind, the centre of the land, the dwelling of majesty. . . . In consequence of high waters, its foundation had become weak, and owing to the filling up of the streets of Babel, the gateway to the palace had become too low. I tore down its walls of dried brick, and laid its cornerstone bare, and reached the depths of the waters. Facing the water, I laid its foundation firmly and raised it mountain high with bitumen and burnt brick. Mighty cedars I caused to be laid down at length for its roofing. . . . For protection, I built two massive walls of asphalt and brick 490 ells[1] [*approximately five hundred and fifty metres*] beyond Nimitti-Bel [*the outer wall of Babel*]. Between them I erected a structure of bricks on which I built my kingly dwelling of asphalt and bricks. This I surrounded with a massive wall of asphalt and burnt bricks.

Bitumen may have found its greatest use in the construction industry, whether that entailed building palaces, ziggurats, bridges, or dams but, as noted earlier, it found wide application in the decorative arts. Bilkadi writes further as follows:

> In the 'Ubaid period, for example, bitumen was moulded into wigs and covered with gold foil or copper to adorn terracotta figurines and large stone statues. It was also used as a base in which to inlay precious stones for vases. And from the Royal Cemetery at Ur, you can see the extent of Sumerian artistic design and craftsmanship in such treasures as the 'Standard', a two-sided panel in which mother-of-pearl figurines are inlaid in bitumen with a mosaic background of lapis lazuli.
>
> This inlay technique was also used to decorate figures in the round – such as a wooden harp with a gold bull's head and a standing goat; the head and legs of the goat, as well as the thicket on which he is leaning, are carved out of wood and covered with gold foil, cemented with bitumen. The back and flanks of the animal were also coated with bitumen in which hair had been embedded.

Practical uses to which the bitumen of Mesopotamia was put are described by other ancient writers as follows:

---

1. Note: 1 ell = 45 inches or 114 centimetres.

> On account of the scarcity of timber their buildings are finished with beams and pillars of palm wood. They wind ropes of twisted reed round the pillars: and then they plaster them and paint them with colours, though they coat the doors with asphalt. (Strabo, XVI)

> There in Babylon, Trajan saw the asphalt with which the walls of Babylon have been built for, together with bricks or gravel, it produces such strength that the walls made of it are stronger than rock and any kind of iron. (Cassius Dio, LXVIII, 27)

> Indian bitumen is valued more than any other; it is considered so excellent because it glistens like unto purple, is heavy and emits a strong odour. The dark and dirty kind, on the other hand, is full of defects; it is mixed with pitch and comes from Phoenicia, Babylon, Zacynthus and Sidon. (Dioscorides 1, 83)

Garstang (1933) reports the discovery of a wall at Jericho enclosing four to five acres of ground. The wall was four feet thick, used bituminous earth as a cementing material, and could be dated to Early Bronze (2500-2100 BC). Layard (1849) makes reference to the walls of Nineveh in similar terms – 'The bricks bonded with asphalt have remained immovable for thousands of years.' Fibrous material was often added to the bitumen to ensure its stability, chopped reeds, rushes, and chaff from threshing being among those most commonly used.

The use of bitumen in a similar manner to that of Mesopotamia extended equally into Egypt, and there is evidence of trade between the two countries. Woolley (1928) writes:

> [T]here are found in Egypt, about the time of the First Dynasty there, which is roughly contemporary with the First Dynasty of Ur, certain products which are unmistakeably borrowed from Mesopotamia, stone mace-heads, cylinder seals, stone vase shapes and the panelled decoration of walls, and such borrowings denote at least a trade connection with the Euphrates valley.

Applications in Warfare

Apart from the applications of bitumen in construction and ornamentation, it had its specific uses in war at a later date. During the Peloponnesian War (431-404 BC) the Spartans used chemical agents during the siege of Plateae (429 BC). They heaped tree branches up against the city walls covering them with tar and sulphur, and set fire. The smoke was so suffocating that the defenders were forced to withdraw from the walls. Fortunately for the Athenian defenders a sudden rainstorm doused the fires. Thucydides, in reporting the siege, says the Athenians 'threw fire with sulphur and pitch' at the enemy.

Aeneas Tacticus (fourth century BC), one of the earliest Greek writers on the art of war claimed a recipe for 'unextinguishable fire':

> Take pitch, sulphur, tow, manna, incense and the scrapings of gummy woods from which torches are made, light the mixture and throw it against the object you wish to reduce to cinders.

It is possible the 'manna' to which Tacticus refers was obtained from *fraxinus ornus*, or the flowering ash, which is native to Sicily, but widely distributed elsewhere throughout the eastern Mediterranean, or to 'turpentine' which may be obtained from a variety of conifers by distillation, e.g. *pistacia terebinthus*, the Chian or Cyprus turpentine tree. Turpentine is an excellent solvent for both sulphur and phosphorus, both of which increase the lethal nature of any mixture and, therefore, a desirable ingredient. The terebinths of Mamre, where Abraham set his tents (Genesis 13:18) after parting with Lot, were a potential source of turpentine, but whether the Canaanites ever used it for such a purpose must remain a subject for speculation.

The 'unextinguishable fire' of Tacticus was probably the predecessor to what later became known as Greek Fire, a weapon of war perfected by the Byzantines around AD 672. Understandably, the recipe was considered a state secret at the time, one which has intrigued later generations. However, it is generally accepted that naptha, a lighter, more liquid hydrocarbon than pitch, was an ingredient vital to its effectiveness, other commodities including sulphur, quicklime, and nitre. Partington (1960) gives a fascinating account of the use of bituminous materials in warfare.

There are no records of the Egyptians having used bitumen, or its allied products, in war. However, Vitruvius (*fl.* c. 75-15 BC), a Roman architect, describes the siege of Apollonia (214 BC) and the actions of an Egyptian, Trypho of Alexandria, who commanded the besieged:

> it was perceived in what quarter their adversaries purposed to make an entrance with their tunnel [*beneath the city walls*]. On learning the direction he filled the bronze buckets with boiling water and pitch overhead where the enemy were, along with human dung and sand roasted to a fiery heat. Then in the night he pierced many openings, and suddenly flooding them, killed all the enemy who were at work there. (Book 10:XVI.10)

It is apparent from the foregoing that bitumen found practical application throughout the entire ancient world, from the very earliest of times by the most primitive of our ancestors through the millennia of Antiquity when Sodom flourished, and extending up to the present day. In fact the commercial exploration and exploitation of the world's hydrocarbon resources over the last one hundred years has led to the growth of an oil industry now dominating the world economy.

Mummification

Much work has been done in recent years analysing the constitution of mummy wrappings and materials used in mummification techniques, funereal practices customarily associated with Ancient Egypt. It has been concluded firstly, that though the 'Bitumen of Judaea' was used in preservation, it was used in association with a wide range of other substances and, in general, the bitumen content never exceeded about 35 percent of the total volume. Secondly, that bitumen was used from other sources also, one of these being identified as the Gebel Zeit (or 'Mountain of Oil') in the Gulf of Suez and, possibly, Hît in Iraq.

Though results of geochemical studies on mummy wrappings indicate the incorporation of conifer resins, greases and beeswax into the bituminous mixture, the analyses have been generally restricted to mummies originating from the Intermediate, Ptolemaic, and Roman periods between 1000 BC and AD 400. This does not discount the possibility of bitumen having been used in earlier Egyptian preservation practices. Nissenbaum (1992) states that existing evidence only dates the use of Dead Sea bitumen in this manner from the Greek conquest of 332 BC. However, there is some evidence that bitumen may have been utilised in mummification as far back as 3000 BC, though this early practice may not have used Dead Sea bitumen. In this connection Wallis Budge (1893) writes as follows:

> The oldest mummy in the world about the date of which there is no doubt, is that of Seker-em-sa-f, son of Pepi I, and elder brother of Pepi II, 3200 BC, which was found at Sakkarah in 1881. . . . The skeletons found in sarcophagi belonging to the first six dynasties fall to dust when air is admitted to them, and they emit a slight smell of bitumen.

Funeral rites were of utmost importance to the Egyptians, and though controversy may exist regarding the extent to which bitumen was used in the preparation of balms for mummification, the practice can be traced back to the earliest dynasties. Such controversy is, perhaps, exemplified by the uncertainty regarding Menes, the first identifiable king of Egypt, who is variably dated as living between 5867 BC and 3623 BC, a span of over two thousand years. Budge declares himself in favour of the chronology of Brugsch-Bey (1879), dating Menes to 4400 BC. Thus, the use of bitumen in embalming procedures possibly dates back to far distant times, long before its general acceptance by archaeologists for that purpose, and proof of its use by scientists. Mummification petered out in Egypt during the Roman occupation, mainly because of Roman disapproval of the practice.

Other scientific investigations, particularly those into the source of bitumen discovered at Tell el Ôoueili in southern Mesopotamia, demonstrate the assistance to archaeologists provided by the forensic chemist (see commentary at the end of this chapter). Analytical techniques have shown the bitumen to have originated from a number of sources, which fit in well with the shifting politics of the region over the time frame involved. Firstly, between 5800 and 4500 BC, bitumen was imported from locations within present day Iran; secondly, between 4500 and 3700 BC, it came from northern Iraq and thirdly, later from local sources such as Hît and Abu Jir. The export of bitumen out of the Dead Sea, during the time when Sodom, Gomorrah and their satellite cities flourished, encouraged an inbound trade of items in return. Archaeological investigations at Bab-edh-Dhra have uncovered evidence of this trade:

> Among the cultural items that reflect foreign contact . . . the majority – including architectural features, cylinder seal impressions, jewelry, some forms of pottery, and a carved bull's head – show Syrian, if not Mesopotamian, influence. (Schaub, 1993)

Transportation of Bitumen

Transportation of bitumen, whether to Egypt or elsewhere, was a long, tedious, and harsh affair following long tortuous routes across semidesert scrubland. Undoubtedly, the caravans would have consisted of trains of asses (or donkeys as now generally known). Though camels are better pack animals than donkeys in many respects, the dromedary is only believed to have been introduced into Egypt during the second millennium, which would have been several centuries later than the envisaged date of Sodom. In any event the ass is likely to have been more capable of coping with the steep, narrow, and wearisome ascents leading out of the Dead Sea. However, camels may have been used latterly for transport across the desert.

There are numerous references to asses, translated from the Hebrew *hamor* or *chamor*, in the Old Testament. Those contained in Genesis are:

(1) Relating to Abraham's gift to Pharaoh when Sarah, his wife, was taken into Pharaoh's house: 'and he had sheep, and oxen, and he asses, and menservants, and maidservants, and she asses, and camels'. (12:16)
(2) In connection with Abraham's intended sacrifice of Isaac: 'And Abraham rose up early in the morning, and saddled his ass, and took two of his young men with him.' (22:3)
(3) As a store of Jacob's wealth; 'And the man increased exceedingly, and had much cattle, and maidservants, and menservants, and camels, and asses.' (30:43)

The mention of asses and camels in the same context has obvious implications in dating the Abrahamic story if, as is considered certain, the camel was introduced into Egypt during the second millennium. Furthermore, there are other references to camels in Genesis, e.g. 24:30-33.

It is inevitable that the bitumen resources of the Dead Sea would have influenced the foreign policy of the Egyptian Pharaohs, ever since the important uses to which bitumen could be put were first realised. The Dead Sea was the closest source of supply to Egypt of a commodity that increasingly became perceived to be both essential and strategic. Any influence the Egyptians exercised over bitumen production may have manifested itself either in the form of direct military control of the region or, what is more likely, the provision of military escorts to trade caravans. Any cooperation by the Egyptians with the Sodomites would have had the common purpose of minimising potential disruptions to peaceful trade, a trade which undoubtedly included a variety of articles and commodities other than bitumen. As this mutually dependent trade prospered so would the consequences of any disruption, and the degree to which military protection was provided would have waxed and waned according to prevailing politics and foreign belligerence. However, the greatest disruption to the supply of bitumen to Egypt would have been the cataclysmic event that obliterated Sodom, Gomorrah, and their satellite cities. And this would have had serious economic and political repercussions.

Trade Routes

Tel Arad, in southern Canaan, was a major centre of trade, being one of the largest cities in the immediate vicinity of the Dead Sea during the Early Bronze Age (Chapter 3). Archaeological evidence has been discovered indicating a network of routes spreading outwards linking Arad to Jerusalem and further north, and to Tel Irani and Palmahim on the Mediterranean coast (see Figure 27). These coastal locations suggest a maritime trade which would have extended southwards into the Nile delta region, and northwards along the Levantine coast. There was also the well known overland route along the coastal plain, known as the 'Way of Horus'. The site of Ein Besor may have been a staging post for trade, as the discovery of a large number of artifacts of Egyptian origin signify strong commercial ties between Canaan and Egypt.

At Tel Arad lumps of bitumen have been discovered, widely distributed throughout the settlement indicating its extensive use in construction, particularly in buildings identified as temples. As a trading centre, much of the bitumen originating from the Dead Sea would have passed through Tel Arad. However, to get there from Sodom was a considerable climb as the difference in elevation between the two locations amounts to approximately one thousand metres, half of which is encountered over a relatively short distance as the path ascends the western escarpment.

Musil (1907) mentions a Roman road, with engineered cut-fill sections, running eastwards from the western shore of the Lisan Peninsula across the Lynch Strait. During an aerial reconnaissance, Glueck (1937) was able to trace parts of other Roman roads, and Harland (1942), quoting Glueck says 'it then probably crossed to the west side of the Dead Sea over the ford which used to be passable from the Lisan to the other side'. In the same work Harland reports that in 1928 Melvin Kyle, in his *Explorations at Sodom*, described a likely crossing of the Lisan Strait, having discovered Roman milestones in the area. Klein (1965) also reports finding a *rujm* (a heap of stones used by Arabs to mark the beginning of a path) on the western shore of the strait (see Figure 5).

There are few tracks branching inland from the western shores of the Dead Sea that attain the plateau beyond the escarpment in the vicinity of the Lynch Strait. The closest to the ford is located about six kilometres to the south, approximately one and a half kilometres north of the town of Ein Boqeq. The track leads up the escarpment almost in a straight line, clinging to the precipitous mountainside for approximately three kilometres, and rising some four hundred metres in the process. From the modern road at its base it appears an awesome ascent. The track is depicted on Figure 34, leading first to Mezad Hatrurim, an ancient fortification situated on a hilltop, and from there winding its way across the rugged plateau to the ancient city of Tel Arad.

This track up the escarpment would have been the shortest route from the western shore of the Dead Sea to Beersheva and beyond, and it is not difficult to visualise long caravans toiling upwards on the first leg of that long journey to the coast, or to connect with the overland route into Egypt through the Wilderness of Shur. A number of wells on the plateau, presumably now defunct, can be observed on large scale topographic maps. The former importance of these wells may be appreciated from the convergence of tracks from all quarters. The most important wells appear to be those at Tel Malhata and Tel Masos, to the west of the modern city of Arad.

Nissenbaum *et al.* (1984) report the findings of archaeological bitumen at Tel Arad and Small Tel Malhata. They relate these artifacts to the trade in Dead Sea

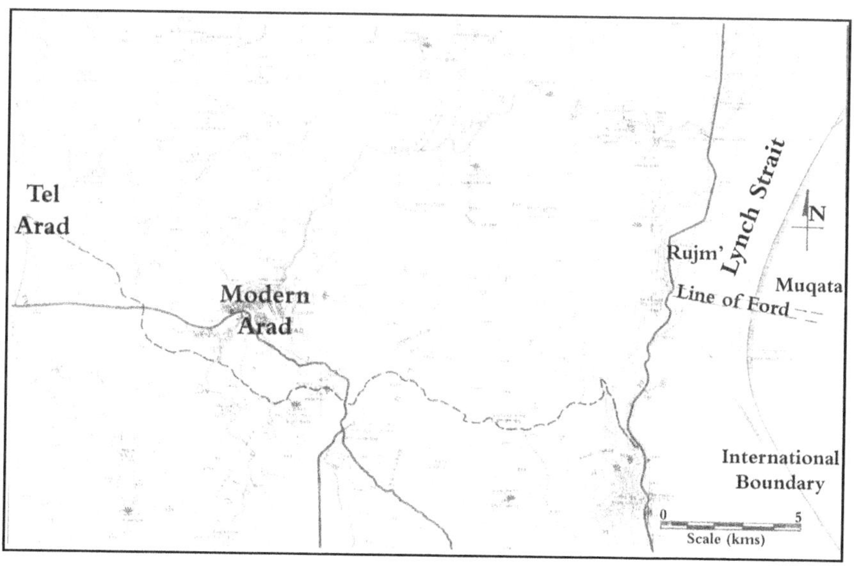

34. The western ascent out of the Dead Sea leading to Tel Arad.

bitumen during EBI and EBII, i.e. between 3200 BC and 2650 BC, indicating the bitumen trade to have extended over a prolonged period of time.

Apart from the torturous western ascent described, there are a number of former routes out of the valley. Less arduous, but no less demanding, would have been the route northwards winding along the western shore to Hazazon-Tamar (or En Gedi) and thence to Jericho, Jerusalem, and beyond; a southerly route leading through the Wilderness of Zin to Kadesh Barnea, and then to Lower Egypt by the Way of Horus; and a route following the Aravah valley to El-Paran, and then by sea to the Red Sea coast of Egypt and thence to Luxor and Upper Egypt.

The eastern shore of the North Basin of the Dead Sea is marked by cliffs that come down precipitously to the water's edge. Thus the route to Damascus and Mesopotamia would have wound eastwards from Zoar through the mountains of Moab to link up with the King's Highway, the route taken by Chederlaomer and his armies when they marched south in their campaign against the Sodomites and their allies.

Connan *et al.* (1992) sum up the trade between Canaan and Egypt as follows:

> The trade between Canaan and Egypt became intensive during a comparatively short period in the Early Bronze Age (Ben-Tor, 1986). In the 150 years from approximately 3250-3100 BC, corresponding to the Early Bronze I period in Canaan and the Naqada III (predynastic or Dynasty 0 period in Egypt), numerous Egyptian wares or Egyptian related materials are found in Canaan.

They refer to a report by Brandel (in Ben-Tor, 1989), where over a thousand vessels of Egyptian origin from Early Bronze I were found at Tel Irani containing evidence of trade in Dead Sea bitumen. They also make reference to sixty clay seals of Egyptian origin of similar age having been discovered at Ein Besor by Porat (1989), with which bitumen was also associated. Both Tel Irani and Ein Besor were known trading centres. Wright (1985) and Ben-Tor (1989) have concluded that Egyptian-Canaanite relations were peaceful, and largely economic.

As evident from the present work, the mainstay of the export trade of Canaan was in the form of bitumen and related products. But other raw materials ancillary to bitumen are likely to have been traded, the production of which will be discussed in the following chapter. There was also a wide variety of agricultural products, as has been pointed out by Tubb (1998), which would have included olive oil, wine and textiles.

The conclusion that may be drawn points to a strong interconnection between the economies of Egypt and Canaan, to the extent they were probably interdependent, perhaps even amalgamated. The disruption of this trade, as a consequence of the sudden and inexplicable obliteration of Sodom and her satellite cities, must be viewed as an event certain to lead to regional instability and economic turmoil.

Demise of the Canaanite Bitumen Trade

One objective in waging war in antiquity was to control trade routes, to monopolise the sources of supply of the commodities involved, and impose taxes. The invasion by the kings Amraphel, Arioch, Chedorlaomer and Tidal from the north, as described in Genesis 14, and discussed in Chapter 1, may be viewed in this light. It is a noteworthy example of disruption to bitumen production in the Vale of Siddim, with its origins rooted in politics. But these earliest of invaders in the historical record, were probably not the first, and certainly not the last.

After the destruction of Sodom (c. 2350 BC) the trade in Dead Sea bitumen with Egypt eventually recovered, though possibly with less centralisation and control than previous. Two thousand years were to elapse before the conquests of Alexander the Great (356-323 BC), and during this period it may be expected that the fortunes of the bitumen industry in the Dead Sea waxed and waned.

Following the death of Alexander, two of his foremost generals, Ptolemy I Soter and Antigonus I Monopthalmos (the One-Eyed), fought each other to gain territorial control of the Middle East. In 312 BC, Antigonus dispatched one of his subordinates, Hieronymus, to conquer the Nabateans and expel them from the Dead Sea region. One of his objectives was to seize control of the bitumen supply to Egypt, then a monopoly of the Nabateans.

The Nabateans were once nomadic peoples, whose oasis settlements bequeathed the name of Nabatene to the desert border lands of Syria and Arabia, and who migrated into the Dead Sea region during the sixth century BC. They prospered by occupying the oases dotting the trade routes through the hinterland. Diodorus Sicilus (c. AD 50) described them as constituting a tribe numbering about ten thousand, who were profitably engaged in far-flung trade extending into Felix Arabia (Yemen) in the east, and Egypt in the west.

A major centre was Petra positioned on a crucial crossroads of trade, and they controlled many of the Red Sea ports, including Elat in the Gulf of Aqaba. At the time of Hieronymus's foray into the Dead Sea the Nabateans were in full control of the harvesting of Dead Sea bitumen. Kuhn (2000) describes the ensuing events:

> When Hieronymus and his army reached the southern end of the Dead Sea, they found many Nabateans camped on the shore. They were waiting for the appearance of what they called the *thawr*, Arabic for bull, in the waters of the Dead Sea. The 'bulls' were big chunks of bitumen which would float to the surface. 'They make ready large bundles of reeds and cast them into the sea. On these not more than three men take their places, two of them row with oars, which are lashed on, but one carries a bow and repels any who sail against them from the other shore, or who venture to interfere with them. When they come near the floating bitumen they jump on it with axes and, just as if it were soft stone, they cut pieces and load them onto the raft, after which they sail back.'

To prepare the bitumen harvested in this manner women and children would sprinkle the chunks of bitumen with sand, presumably to prevent them sticking together, and the pack them into leather bags on camels. The camel caravans would then head in a southeast direction towards Petra before swinging northwest for Gaza and the coast. From there the bitumen was taken either by land or sea to the Nile delta.

Two observations may be made from Kuhn's account. Firstly, the fact that precaution had to be taken against any interference with the harvesting of the bitumen, especially by those 'from the other shore', suggesting that the residents of the Dead Sea were divided into factions at the time, and the hostility was territorial. It is likely 'the other shore' was the western shore, and that relations between the Nabateans and the inhabitants of Judaea were not always peaceful. Secondly, for the camel caravans to wend their way towards Petra, before heading for Gaza, supports that view, as it indicates the western ascent out of the Dead Sea, and other routes across Judaea were not available to the Nabateans.

Hieronymus's attempt to take control of Dead Sea bitumen production was a vain enterprise, for the Nabateans sent out an army of six thousand men to engage in battle in order to defend their territory. Further attempts by Antigonus (the One-Eyed) to conquer them, and blockade the export of bitumen to Egypt, also failed. The Macedonians attacked Petra, but found its fortifications too well built to overcome, and the surrounding desert hostile. They eventually negotiated a withdrawal, Ptolemy I Soter solidifying his hold on Egypt, and founding a dynasty. During his reign the world-famous library of Alexandria was established.

For the next two centuries the Nabateans conducted a lucrative trade with Egypt, but this trade was, on occasions, interrupted by discord and hostility. King Alexander Jannaeus, who ruled Judaea between 103-76 BC, was no more than a cruel and bloody tyrant, and his kingdom was in constant conflict during his twenty-seven year reign. He waged war upon the Nabateans, as he waged war on almost all his neighbours. It is reported that on one occasion, acting as High Priest, he stirred up the emotions of his own people to such an extent that he could subsequently justify his soldiers slaughtering six thousand of them. At the conclusion of a civil war that lasted six years, in which fifty thousand Judaeans died, he crucified eight hundred rebels, and had the throats of their wives and children slit while consorting with his concubines.

In 88-87 BC, the Seleucid king, Antiochus XII, ventured into Nabatean territory in an attempt to gain control of the bitumen industry, but failed dying in battle. Eventually superior forces of the Romans came, and annexed the Nabatean kingdom. It is said that the Roman general, Marc Antony, was so infatuated with Cleopatra, Queen of Egypt, that he gave her the Nabatean 'oil fishery' as a gift. She, with considerable business acumen, leased it back to the Nabateans on payment of two hundred talents a year. Estimated as about $400,000/year in today's valuations, this was an enormous sum. Eventually the

Nabateans revolted, refusing to pay this annual tribute, and the Romans sent out an expeditionary force against them under King Herod. The Nabateans defeated them at the Battle of Qanawatt in present-day Syria. The following year the fleets of Marc Antony and Cleopatra were defeated at the Battle of Actium.

The importance of bitumen to the politics of the Middle East cannot be over-estimated. In his work on the molecular archaeology of ancient bitumens Connan (1999) refers to this period:

> Bitumen from the Dead Sea was exported to Egypt from the fourth-third millennium BC (Connan et al, 1992). Obviously, this fossil product was still strategic for Egypt at the end of the Ptolemaic period. For example, in 30 BC Cleopatra lost her entire fleet in Alexandria during conflict with the Nabateans from Petra. They were extremely angry with the Queen of Egypt who put an end to their monopoly on the exploitation of Dead Sea bitumen and consequently one of their sources of prosperity. They organized a raid on Alexandria and set fire to Cleopatra's fleet, which was completely burnt.

Commentary on Modern Petroleum Chemistry

The remains of all living matter, including by-products such as fats and waxes, provide the diagnostic chemist with an abundance of organic materials on which to apply his analytical expertise. As a result we are now in a position to identify more accurately than hitherto the true nature and molecular composition of this unique range of substances, especially those falling within the classification of hydrocarbons.

The most powerful tools include gas chromatography, mass spectrometry and isotope chemistry, which also have considerable application in other fields, such as forensics, e.g. drug identification, and pollution control. Of the most interesting is their considerable potential in archaeology.

Gas chromatography is one of the specialist techniques that enables the volatile and semi-volatile components of an organic mixture to be separated for identification, without altering the molecular structures of the individual constituents. Two gas chromatograms are given in Figure 35, showing an example for a typical crude oil, and a resultant gasoline product. The various ranges of product resulting from the fractional distillation of the crude oil are indicated, though there is much variation in practice. Naturally, the higher the carbon content the heavier the distilled products.

Difficulties often arise in interpreting the chemistry of older bitumens because of the inevitable changes that take place over extended periods of time, i.e. from atmospheric weathering and bacterial action which, understandably, can seriously alter the molecular structure of the original material. The first objective of the geochemist, therefore, is to identify those molecular structures most resistant to time-dependent change. These are known as biomarkers, and constitute the chemical 'fingerprint' of a substance. In the field of archaeological bitumens the most useful are the steranes and the terpanes. These are isomeric

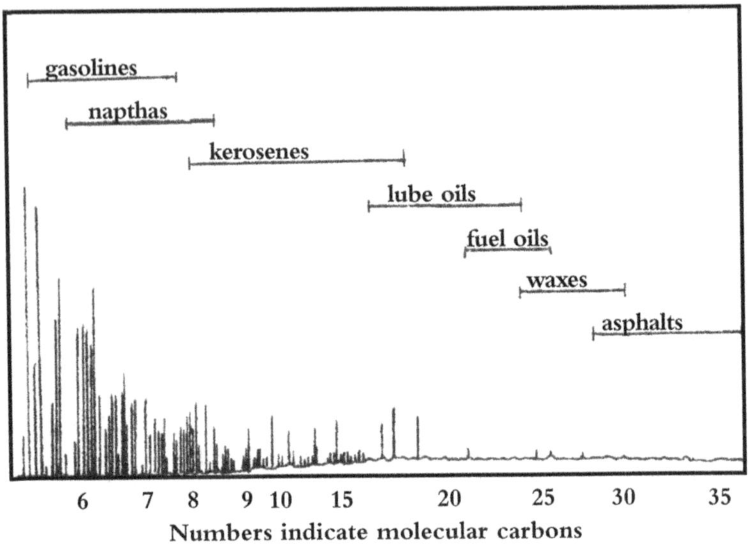

(a) Crude oil

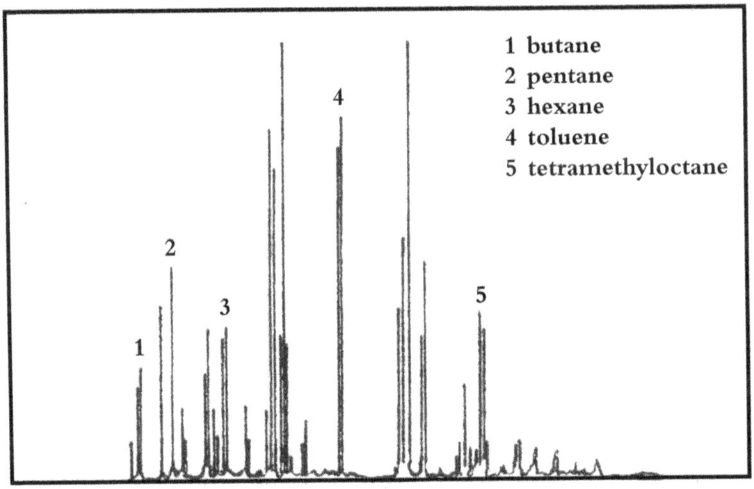

(b) Gasoline

35. Comparative gas chromatograms for a typical (a) crude oil and (b) gasoline.

saturated hydrocarbons possessing complex molecular structures, the proper name for sterane being perhydrocyclopentanophenanthrene! These have become the targets of the analytical chemist attempting to identify sources of archaeological bitumen, and comparing them with those available in the data bank.

Connan (1999) reports that over seven hundred bituminous samples have been analysed using such advanced techniques on specimens drawn from a wide geographic region – Iraq, Syria, Iran, Bahrain and Kuwait. The results of this work provide extensive and reliable reference material to enable the identification of the sources of bitumen encountered in archaeological materials generally, with a far greater degree of confidence than previously. Because bitumen was carried along trade routes from source to market, it has been possible to determine some of the caravan routes followed by traders in the wider Middle East. The similarity is akin to the techniques used by forensic criminologists in analysing samples of human tissue for DNA, and linking the results to potential suspects.

Gas chromatograms are reproduced in Figures 36 (a) and 36 (b) demonstrating the similarities between two basic bitumens found in the Dead Sea region, as obtained from natural sources linking them to archaeological sites, bearing in mind that over the intervening millennia the bitumens have degraded somewhat with loss of their lighter components. The composition of the two main types of Dead Sea bitumen, occurring (a) in Dead Sea floating blocks and (b) outcrops at Nahal Heimar, show striking similarity with those specimens recovered from Maadi (Egypt) and Palmahim respectively on the Mediterranean coast.

With regard to Dead Sea bitumen, Connan *et al.* (1992) summarise their findings in the following words:

> Lumps of archaeological bitumens, found in excavations of Canaan and Egypt, dated 3900-2200 BC, have been genetically linked to natural deposits of the Dead Sea area. One of them, the Tel Irani bitumen, Early Bronze I in age, has been shown to be identical to floating block Dead Sea asphalts. All other archaeological bitumens, biograded to various degrees, have been recognized as floating block asphalts that have been subsequently weathered at archaeological sites.

Furthermore, they conclude:

> The data presented herein provide the first evidence of export and trade of raw bitumen from Canaan not only to Egypt (Maadi) but also to Egyptian trading posts (Ein Besor, Tel Irani, etc.) on the route to mainland Egypt as early as 3900 BC.

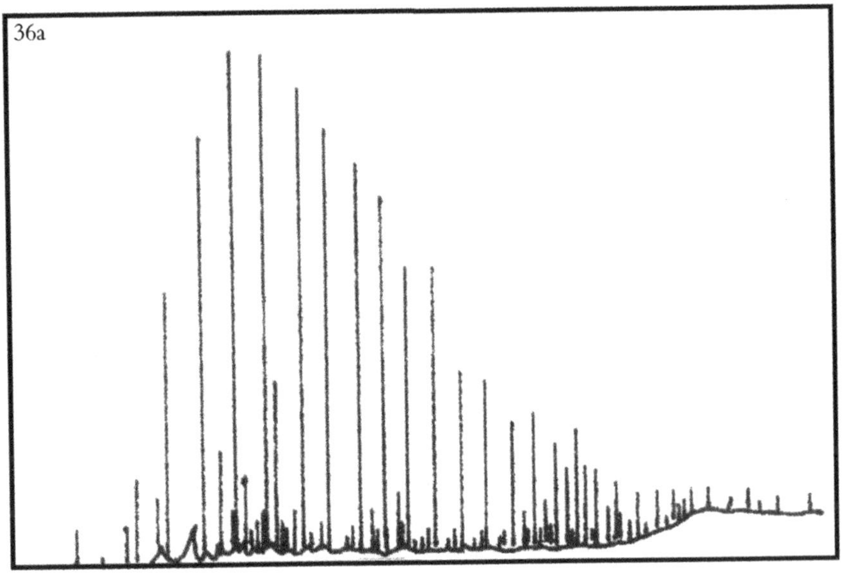

Dead Sea Floating Block

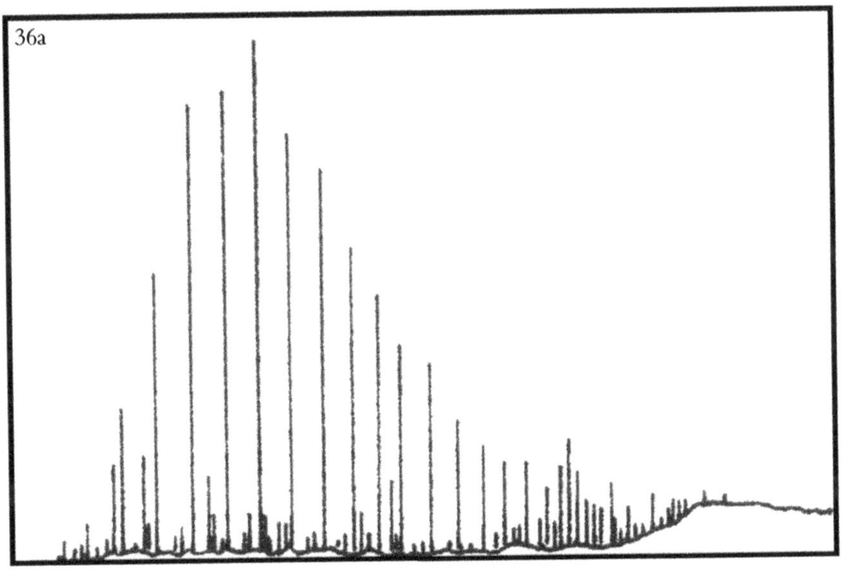

Maadi (Egypt)

36. (a) Similarities of bitumen from the Dead Sea and Maadi, Egypt (after Connan et al, 1992).

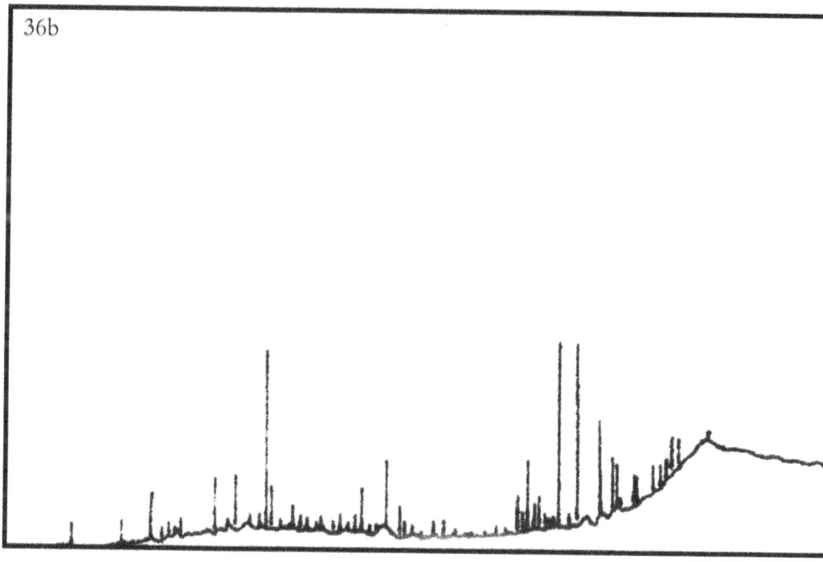

Nahal Heimar

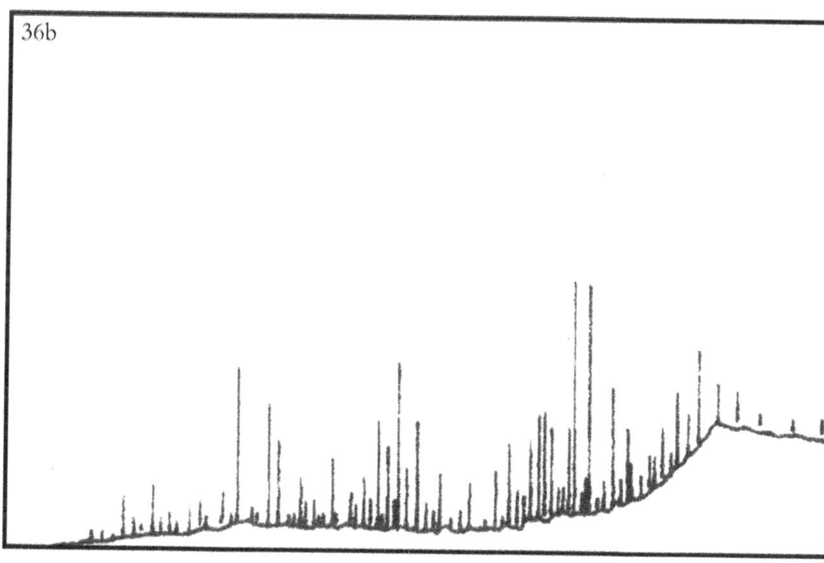

Palmahin

36. (b) Similarities of bitumen from Nahal Heimar and Palahim (after Connan et al, 1992).

6. Mineral Resources in Antiquity

Regional Resources

Aside from bitumen, what other mineral resources were available to the Sodomites which would have enhanced their commerce? One of the most prominent topographical features is, of course, Mount Sedom in the southwest quarter of the South Basin. This is a mountain of relatively pure salt (*sodium chloride*). There is, also, an abundance of limestone (*calcium carbonate*) and silica, and the salt flats of the South Basin which include impure evaporates, mostly salts of potassium and magnesium, such as *carnallite* and *sylvinite*, with varying amounts of *sulphates* and *bromides*. All could have been used as the foundations of a chemical industry, albeit a primitive one, located around the southern shores of the Dead Sea.

The term 'chemistry' is generally believed to originate from *ch meia*, an ancient name for Egypt, though an opinion expressed by Davidovits and Morris (1988) suggests the original root is *khem* or *Khnum*, one of the most ancient of prehistoric Egyptian gods. The Famine Stele, discovered on the island of Sehel, near Elephantin, in 1889, was erected during the reign of Ptolemy V Epiphanes (205-182 BC), and from its 2600 hieroglyphics almost one-third relate to rocks, mineral ores, and their processing. This indicates the importance the Egyptians placed upon the art and practice of alchemy.

The abilities of the Egyptians in the fields of alchemy, or chemistry as it is better known today, were highly respected and often feared. In AD 296 the Emperor Diocletian ordered the destruction of all chemical texts, having a baseless fear that the possessors of this 'black art' might transform base metals into gold and, thereby, undermine the economy of the Roman Empire. His act is regrettable, as we have little knowledge of the level of skills the ancient alchemists attained. The belief is that they were far more advanced than we generally think. It is reasonable to assume that with the bitumen resources of the Sodomites, so vital to the commercial well-being of Egypt, much of that proficiency in chemistry would have migrated into the Dead Sea valley.

In terms of the available mineral resources of the Dead Sea, it is not unlikely that the Sodomites knew how to calcine limestones to produce quicklime, to slake it with salt to produce caustic soda, an intermediate stage in the production of soda (or *natron*, otherwise known as sodium carbonate) which, in turn, has numerous applications in glass manufacture, glazing and faience work, and for the preparation of *sodium silicates*.

The Dead Sea valley is a hot, unpleasant place, probably as hot and unpleasant now as it was over four thousand years ago. Without doubt it was a cruel, harsh existence, not only for the Sodomites themselves, but for those forced to labour beneath them, and they would have needed much manpower. The life of such labourers, toiling in that ancient 'hive of industry', would not have been little worse than those of workers in a modern sweatshop.

Today the South Basin of the Dead Sea, on both the Jordanian and Israeli sides, is one vast industrial complex, producing a wide variety of chemical products from the waters of the Dead Sea. These chemicals include not only fertilisers such as potash, but bromine and magnesium, the latter produced by the electrolysis of magnesium chloride. The ancient Sodomites truly would have marvelled at the wonders of present day technology!

Let us take the opportunity in this chapter to indulge in the supposition that the Sodomites did acquire practical knowledge, based on experimentation and observation, two prerequisites of the modern day scientist, to enhance the raw materials so readily available to them, and to exploit this knowledge in the interests of trade. The basic chemistry involved is little more than 'schoolboy stuff', and modern symbols have been used to explain the chemistry of the processes involved.

Sulphur in the Dead Sea

Genesis (19:24) tells how 'the Lord rained upon Sodom and Gomorrah brimstone and fire from the Lord out of heaven'. However, brimstone, more commonly called sulphur, is a substance found in relative abundance at certain locations within the valley of the Dead Sea. It has been claimed that 'a nation's consumption of sulphur is a measure of its industrial progress', and this bold statement may have been as true in the days of Sodom as it is today. Since sulphur is an element of considerable importance as a feedstock to the chemical industry of today, undoubtedly that importance would have extended back into Antiquity.

Zeng *et al.* (2011), in their research into the genesis of elemental sulphur in the natural environment, recovered a number of sulphur balls from an offshore hydrothermal environment at Kuishantao, Taiwan. Some of these balls attained diameters of about six centimetres, and weighed 300 grams. The balls showed evidence of having accreted from nuclei. They summarised their findings as follows:

> Native sulfur is mainly found in three geological environments: hydrocarbon-bearing basins, volcanic settings, and sea-floor hydrothermal settings. In the hydrocarbon-bearing basins, native sulfur deposits can be classified as bioepigenetic native sulfur deposits that result from the biogenetic, post-sedimentary conversion of anhydrite or gypsum into hydrogen sulfide, that is subsequently oxidized to elemental sulfur (e.g Jassim *et al.*, 1999).

> Such deposits occur mainly in cap rocks over salt diapirs and in stratabound deposits in bedded evaporates (Alonso-Alcárate *et al.*, 2001). Biogenetic syngenetic sedimentary native sulfur deposits result from bacterial reduction of SO_2 [*sulfur dioxide*] to H_2S [*hydrogen sulfide*] to native sulfur by oxygenated surface waters in a lagoonal or otherwise restricted evaporative or euxinic marine and lacustrine environment (e.g. Lein *et al.*, 1975).

Work by Torfstein *et al.* (2008) at the Institute of Earth Sciences, Jerusalem, concluded freshwater sulphates in the Dead Sea basin to be the prime origin of the significant amount of sulphur found embedded within the Lisan marls and associated sediments. Through bacterial action within the anoxic water and lake bottom sediments, the sulphates are reduced, firstly to sulphides, and then to the elemental sulphur. *Beggiatoa*, which consists of slender unbranched filaments, is one of the better known bacteria which utilise hydrogen sulphide as a source of energy, and this appears to be instrumental in the reduction process, though other bacteria of which we have little knowledge, may equally be involved. The resulting elemental sulphur is stored in the cells of the bacteria after conversion, and when the bacteria die the residual sulphur becomes entombed amidst the sediments.

The Dead Sea environment meets all the criteria noted by Zeng *et al.* for the development of sulphur balls, possessing no shortage of sulphur to fuel the bacterial processes involved. Huge thicknesses of evaporates in the form of interbedded aragonite, calcite, and gypsum are present, the latter containing approximately 19 percent sulphur. Gypsiferous sources, as well as those sulphur contaminants associated with the hydrocarbons issuing from the host rocks, ensure a continuous supply of sulphur compounds for potential conversion. Furthermore, elemental sulphur is frequently encountered during oil exploration in the proximity of intrusions of rock salt, and plug-like formations of rock salt, or diapirs. As noted in Chapter 2 these occur within the Dead Sea region, the most famous of which is Mount Sedom.

Though the thinking is not universally accepted, the wide body of evidence available suggests much of the sulphur found in the form of nodules in and around the Lisan Peninsula, and elsewhere in the Dead Sea basin, can be attributed to anaerobic bacteria as explained. Nodules of sulphur may be observed in large numbers within the marl formations of the Dead Sea, often forming prominent layers within the calcareous sediments. The layered aspect of such nodules suggest a contemporaneous genesis of the individual formations, such as that which might be envisaged by concretions in areas where bacterial activity is concentrated.

Ron Wyatt, a Biblical archaeologist, who examined certain sites along the western shore of the Dead Sea during the 1980s, in pursuit of the remains of the legendary cities, reported finding conclusive evidence of the brimstone that 'the Lord rained upon Sodom and Gomorrah'. He describes on his amateur videos how he found remains of sulphur balls randomly distributed within certain marl formations in the vicinity of Masada, describing them as irregular in size, partially vitrified, possessing an aureole of ash, and showing signs of incineration. An analysis of the sulphur balls indicated them to have a purity of 95.72 percent. It is well known that sulphur balls are often among the ejecta thrown out of erupting volcanoes. However, in this particular instance it is difficult to suggest a volcanic source from where Wyatt's sulphur balls may have originated. An impartial conclusion suggests the role of anaerobic bacteria in their formation would make better science.

Processing of Sulphur

The most primitive technique to improve the purity of native sulphur deposits contaminated with materials such as gypsum, detrital quartz, and limestone, is to melt them. A large heap, which may contain up to several hundred tons, is built upon sloping ground and covered with the residue from a previous melting. Clay is added to form an air-tight roof, with a central air-hole or vent at the top of the stack. The heap is then ignited, and the lower opening of this improvised furnace almost closed to ensure combustion proceeds slowly.

Sulphur is quick to light and burns with a blue flame. Some of the sulphur is consumed, producing sulphur dioxide, but the greater part melts and trickles down into collection moulds at the lowest point of the stack. The refining process can take up to three months depending on the size of the heap being treated. A significant amount of sulphur dioxide is emitted during the process, and this can cause environmental havoc as it is a colourless, poisonous gas of considerable pungency. With a system of collection pipes the gas may be passed through water to form weak sulphurous acid, an intermediate stage in the production of sulphuric acid.

Sulphurous acid is somewhat unstable. If left for any great length of time its potency will be reduced as the sulphur dioxide gas comes out of solution. The chemical reactions are as follows:

$$S + O_2 \rightarrow SO_2$$
$$H_2O + SO_2 \rightarrow H_2SO_3$$

Sulphurous acid, even in a fairly weak condition, has a considerable number of applications. It can be used as a bleaching agent for textiles and woven materials, it possesses antiseptic properties useful in medicine and surgery, and it can arrest or help control the fermentation process during the manufacture of wines and ales. As a gas, sulphur dioxide is often used as a fumigant.

The burning of sulphur in air also produces a small amount of sulphur trioxide, the proportion of which is far exceeded by that of sulphur dioxide. Sulphur trioxide is also soluble in water to form sulphuric acid:

$$2S + 3O_2 \rightarrow 2SO_3$$
$$H_2O + SO_3 \rightarrow H_2SO_4$$

It is doubtful whether those early chemical engineers in Sodom would have recognised the distinction of the sulphurous or sulphuric forms of the acid, but what they would have soon realised from any experimentation is that the aqueous solution of the gas arising from the combustion process had a number of very beneficial applications.

Because of the simplicity in purifying sulphur there is no reason to doubt the possibility of sulphur forming part of the export trade of the Sodomites.

Caustics

The desirability of utilising various caustics in the enhancement of the bitumen refining process has been mentioned in Chapter 4, notably caustic soda (NaOH) and natron, or sodium carbonate ($Na_2CO_3.10H_2O$). The latter material occurs where an evaporitic climate has prevailed for extended periods of time in a highly alkaline sodium rich environment. The formation of natron in nature may be represented by a two stage process. The highly alkaline sodium rich water firstly absorbs carbon dioxide from the air to produce sodium bicarbonate ($NaHCO_3$), which then reacts with more alkali to produce natron. This is represented by the following equations:

$$NaOH + CO_2 \rightarrow NaHCO_3$$
$$NaHCO_3 + NaOH \rightarrow Na_2CO_3 + H_2O$$

A Coptic papyrus of the ninth to tenth century mentions soda, referring to it as *nitron*, the Hebrew term being *neter*. The Egyptian soda lakes were called *Νιτρίαι* by Strabo. There is little doubt that both these terms refer to soda (*sodium carbonate*).

The main source of soda in Egypt was at Wādī al-Natrūn near Memphis, from which soda is still extracted. Other travellers and historiographers who wrote of the natron beds of Egypt, apart from Strabo, included Herodotus (484-425 BC), Pliny (AD 23-79), and St. Jerome (AD 347-420). Pliny noted that 'Egypt had no outcrops of natron except for those near Naucratis and Memphis, the products of Memphis being reputedly inferior.'

Soda, or natron, was used widely by the Egyptians, in the home, in commerce, and in mummification and, by virtue of the advanced knowledge of chemistry by the Egyptian alchemists, it must be expected that the Egyptians were well-versed in the properties and applications of natron. A variety of soaps and cleaning products may be created by blending natron with various oils. Since it possesses antiseptic properties it had wide applications in medicine, to preserve fish and meat, as an insecticide, and for bleaching. It was a substance of great utility in the commerce of the day. Though occurring in nature it can be manufactured.

Salt and the Manufacture of Natron

It has often been assumed that salt from Mount Sedom was used primarily as a food additive, and may have been exported to Egypt. This is possible, though rock salt is notably deficient in iodine, an important trace mineral essential to the healthy functioning of the thyroid, goitre being a symptom of iodine deficiency. However, iodine is present in sea salt, and it is difficult to believe the physicians of Egypt, and elsewhere in the ancient world, would not have been aware of the superiority of sea salt as a food additive in contrast to the purer rock salt. Egypt would have had a sufficiency of sea salt from sources along the Mediterranean or Red Sea coasts.

An interesting possibility arises in considering rock salt from Mount Sedom as a feedstock for the manufacture of natron. Salt mined from Mount Sedom would have been relatively pure in comparison to salt obtained directly from the Dead Sea, which would have been adulterated by a variety of compounds.

The stepwise chemistry involved in the artificial production of natron, though representing a crude process, might well have been along the following lines:

(1) Calcine raw limestone rock, i.e. heat under controlled conditions, to produce quicklime.
$CaCO_3 \rightarrow CaO + CO_2$
The carbon dioxide is driven off.

(2) Slake the resultant quicklime (CaO) with a concentrated, or saturated, solution of brine to produce slaked lime.
$2CaO + H_2O + 2NaCl \rightarrow Ca(OH)_2 \downarrow + CaCl_2 + Na_2O$
Slaked lime is insoluble and will precipitate. Sodium oxide being unstable immediately reacts with the excess water to produce caustic soda.
$Na_2O + H_2O \rightarrow 2NaOH + heat$

(3) Remaining is an aqueous solution of calcium chloride and caustic soda. As the solution cools, carbon dioxide from the air reacts with the caustic soda to produce sodium bicarbonate, which crystallises out as evaporation proceeds.
$NaOH + CO_2 \rightarrow NaHCO_3 \downarrow$
The calcium chloride rich liquor is drained off.

(4) The final stage of the process is to heat the sodium bicarbonate to drive off carbon dioxide to produce natron (or soda).
$2NaHCO_3 + heat \rightarrow Na_2CO_3 + H_2O + CO_2$

Soap

An important use of natron was in the manufacture of soap and cleansing agents. The earliest evidence of soap manufacture in the region is found in Ur III, a Mesopotamian dynasty variously estimated as flourishing within the period 2400-2000 BC. Although there is no specific reference to soap, recipes mention the boiling of oils, such as palm oil and castor oil, with various sodas, salts and resins. It is clear from the context that soap was the end product, though perhaps not in a form easily recognisable today. Also, there is mention in ancient texts of soda and vegetable oils being mixed with urine, fuller's earth, or ashes. All this entailed knowledge of elementary chemistry, knowledge that extended into everyday life.

The Egyptians bathed frequently for two main reasons – hygiene and magico-religious – though it was only the rich who possessed baths. This was in contrast to the Mesopotamians who, in general, only washed their entire bodies on festive occasions for they viewed purity of mind of greater importance than cleanliness of body. Furthermore, the Egyptians, especially women of wealthy families, used a variety of unguents, creams, pomades, rouges, powders and lacquers for adornment, and this implies some knowledge of the chemical properties of the ingredients involved.

Glass

One major use of natron was in the manufacture of glass, bottles being made by winding hot plastic rods of glass around a sand core, and then heating to fusion. Archaeological evidence suggests glass was first produced on the Mediterranean coast in what is now Syria, and later introduced into the Old Kingdom of Egypt (c. 2700-2100 BC). A number of glass objects have been found in Mesopotamian cemeteries also dating to Ur III (2100 BC). The Egyptians do not appear to have manufactured much glass themselves until after about 1500 BC. However, Forbes (1964/72 vol V) claims pottery glazes were used extensively during the fourth millennium both in Egypt and in Mesopotamia, the earliest known glass objects appearing to be faience beads from the third millennium. Though the Late Bronze Age was to witness a rapid growth in glass manufacture, it is a matter of speculation as to what extent the artisans of Sodom may have participated.

Substance	*Formula*	*Ancient Egyptian*	*Modern*
Silica	SiO_2	55-65 %	65-75 %
Lime	CaO	3-10	5-1
Magnesium oxide	MgO	3-5	>1
Alumina	Al_2O_3	1-3	1-2
Iron oxide	Fe_2O_3	1-2	>1
Potassium monoxide	K_2O	1-3	ca 1
Sodium monoxide	Na_2O	15-22	12-15

Table D – Comparative chemical analysis of Ancient Egyptian and modern glasses (after Forbes 1964/72 vol V)

Forbes tables the analytical composition of a glass from Ancient Egypt in comparison to that of modern glass (see Table D). It will be noted that modern

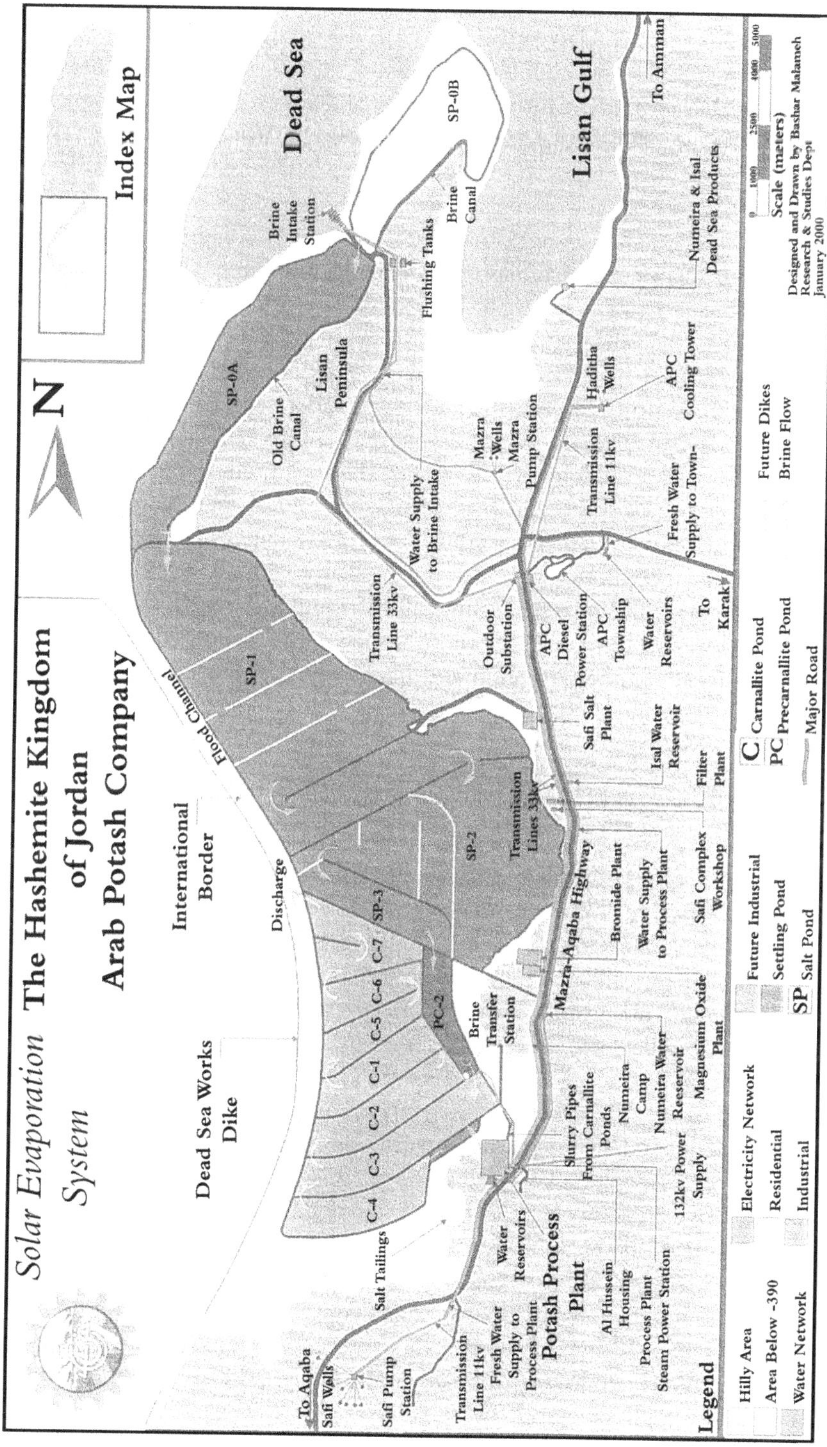

37. Schematic layout of the evaporation ponds of the Arab Potash Company, Jordan.

glass contains a higher percentage of silica, otherwise there is a remarkable similarity in composition. Glass can be manufactured from melting pure silica sand (SiO_2), but this requires very high temperatures of 2300°C. Such temperatures are readily obtainable in modern furnaces, thereby attaining a higher silica content in modern glass compared to that of ancient glass. However, the melting point of silica can be markedly reduced to about 1500°C by the addition of natron or soda (Na_2CO_3), the resultant 'soda glass' being softer as a result. To render it more durable, oxides of calcium (CaO), magnesium (MgO) and aluminium (Al_2O_3) are required in small quantities, and this is equally applicable whether the glass is ancient or modern.

Forbes noted that the natron beds of Egypt are deficient in potassium. This deficiency must have been overcome by Egyptian glass-technologists, as is obvious from Table D, being sufficiently skilled in their art to increase the content of soda ash in order to utilise a lower fusion, and increase the lime content to offset the consequent softening of the final product through adding small quantities of potassium and magnesium. This poses the question – from whence did they obtain their potassium (K_2O)? Carnallite ($KCl.MgCl_2.6H_2O$) is a second-stage evaporation product obtained from Dead Sea water. It is a double salt of potassium and magnesium chlorides, and is crystalline in form. The possibility exists, therefore, that carnallite may have been utilised in Egyptian glass-making to adjust the potassium and magnesium contents to the required levels in order to produce quality glass.

Dead Sea Carnallite

Figure 37 shows a schematic plan of the sequential evaporation ponds operated by the Arab Potash Company in the South Basin of the Dead Sea. It will be noted how the ponds abut the border with Israel, and wrap themselves closely around the Lisan Peninsula to maximise the limited availability of Jordanian space. It may be noted that a new pond, proposed at the time the plan was prepared, extended northwards into the area postulated to have constituted the ancient Vale of Siddim before its loss, referred to as the Lisan Gulf on the map. One section of this dyke complex is reported as having experienced instability during construction.

Dead Sea water is drawn from the North Basin into the first pond, where the first stage of the evaporation process begins to concentrate the solids in solution by removal of its sodium chloride by precipitation. The enriched liquor then flows into successive ponds until the salt has been removed by crystallisation and precipitation. A continuation of the evaporation process in other ponds begins to result in the production of carnallite, or potassium magnesium chloride, and other salts, which constitute the feedstock for further refining and manufacture of chemicals, and which are harvested by dredging. Adler and Harris (1983) outline some of the unique physical properties which complicate the handling of carnallite. A common sight is that of flocs of salt floating upon the ponds (Figure 38).

38. Salt flocs upon an evaporation pond.

It is not impossible that the simple process of evaporation, using a sequence of ponds to control stage deposition of various chemical precipitates, was known to the Sodomites. Whether or not they exploited it for the manufacture of carnallite, to be used for a multitude of applications, not necessarily restricted to glass manufacture, must remain a matter for further study.

Carnallite in Pyramid Construction

One of the more contentious issues raised over the last few decades, is the composition of some of the materials used in the construction of the pyramids of Ancient Egypt. Joseph Davidovits, a French materials scientist, first voiced his opinions about the Great Pyramid of Giza, claiming much of it was not built from cut stone, but utilised a form of reconstituted limestone similar to the concrete of today. As might be expected such radical views did not receive universal acclaim, especially from mainstream Egyptologists. His views were first published in France (Davidovits, 1983), and later (Davidovits and Morris, 1988).

Davidovits was convinced the Egyptians possessed neither the tools nor the technology to carve, haul, and raise limestone blocks weighing from 22 to 15 tons. The tools they possessed were of copper and bronze, too soft for accurate chiselling, and the carving and hauling would have left a litter of broken stone blocks and fragments of unused masonry. Furthermore, the work would have required an immense labour force over a prolonged period. His convictions were supported by the appearance of many of the supposedly carved blocks. They were too uniform in size; they gave the appearance of having been cast in forms, exhibited all the surface defects of modern day concrete, e.g. air holes and segregation; and 'mortar' joints existed between upper and lower blocks that gave all the semblance of the upper block having been cast before the lower block had fully hardened. Also, the joint spacings between adjacent casing blocks were as small as 0.002 inches, an impossibility for carved stone of natural origin. The most convincing evidence, however, was the fact that microscopic examination showed the limestone fossils within the blocks to be fragmented and dispersed randomly, not stratified as would be the case in natural limestones.

The chemistry involved in the manufacture of artificial stone having the appearance of limestone has been determined by Davidovits (2008). In connection with the Giza Pyramid two separate chemical processes have been shown to be valid, by which both core blocks and casing stones may have been manufactured. As these are dependent on the local availability of clay containing an abundance of alumino-silicates, and the chemical composition of clay can be highly variable, the specific technology used at individual pyramid sites may be expected to vary depending on the nature and availability of appropriate source materials. However, in his own laboratory a crew of ten men built a structure consisting of fourteen blocks, each weighing between 1.3 and 4.5 tons, in a matter of a couple of days, thus confirming the ability to construct artificial stone identical in appearance to pyramid blocks both quickly and efficiently. Thus, there is sufficient theory confirmed by demonstration to support Davidovits's views on this issue, which ought to be given greater credence, not only by Egyptologists but archaeologists in general.

In the case of the Giza Pyramid there is a wadi to the south where a relatively soft limestone has been quarried. The limestone contains a higher proportion of kaolinite, an alumino-silicate with the chemical formula $Al_2Si_2O_5(OH)_4$, alumino-silicates being a prime constituent of modern concrete. Davidovits expressed the opinion that this wadi was the likely source of the aggregate used in the construction for much of the Giza Pyramid. It would have been a simple matter for the soft limestone to have been broken up, slurried with natron and carnallite in huge ponds, agitated by stirring to form a homogeneous mix, then allowed to evaporate to a consistency that the product could be conveniently handled. The product would then have been transported, lifted and placed in position in formwork until hardening occurred and the hydration process was complete. The entire process would be akin to the setting of concrete. The rudimentary chemistry of the process involves the inter-reaction of clay ($Al_2Si_2O_5(OH)_4$), natron (Na_2CO_3), lime ($Ca(OH)_2$) and carnallite ($KCl.MgCl_2.6H_2O$), the carnallite being needed to control alkalinity.

The introduction of carnallite into the envisaged process is intriguing for it suggests a possible link to the Dead Sea, and the postulated prowess of the Sodomites in the field of chemical engineering. As described already, carnallite is a product obtained during an intermediate stage in the evaporation of Dead Sea water, and a raw material for the manufacture of potash ($K_2CO_3.2H_2O$) in modern chemical plants today. Though carnallite is deliquescent, i.e. dissolves in moisture extracted from the air, deliquescence does not occur significantly with a relative humidity below 60-70 percent. Therefore, it could have been transported safely through the arid regions extending between the Dead Sea and Egypt. Its usage in the process envisaged by Davidovits and his co-workers would result in traces of residual salt, or sodium chloride (NaCl), remaining within the 'concrete' and this is reported in the analyses of the pyramid stones.

Davidovits and Morris (1988) state that the oldest known remains of high quality cement, or artificial stone, are 9000 years old and can be found amidst the ruins of Jericho. Elsewhere, mortars of similar composition have been found in Syria and Turkey, which appear to be contemporary. They concluded that the manufacture and use of geopolymerised concrete, as it is now known, likely originated elsewhere, being imported into Egypt, not only for use in pyramid construction, but for numerous other applications. The Dead Sea region could well have been a source of this innovative technology.

Other workers have validated Davidovits's original thesis using advanced analytical techniques, such as electron microscopy and X-ray diffraction. Barsoum *et al.* (2006) are among the foremost of these. The conclusion of this work was that Davidovits's thesis regarding the use of reconstituted 'concrete' in the construction of the pyramids may not answer all the questions, but it is a plausible thesis that goes a long way to obviating many of the perceived difficulties confronting researchers into the methodology used by the Egyptians in the construction of the pyramids. Credit must be given to the pioneering efforts of Davidovits and his co-workers, in introducing a new approach to the resolution of that age-old enigma regarding how the pyramids were built. In the meantime the arguments can be expected to rage.

To support his thesis, Davidovits and his co-workers cast a large number of items from geopolymerised concrete. These are reported to have confounded geologists, who on seeing them for the first time asserted them to be carved from natural stone. There are also artifacts on display in various museums dating from Antiquity, such as vases and bowls of extraordinary thinness and intricacy. The common assumption these were carved from natural stone, using copper chisels, ought to be questioned in the light of Davidovits's research.

Silicates

An interesting question to ask, is whether silicates, especially sodium silicate (Na_2SiO_3) could have been produced by those early chemical engineers of Sodom? Sodium silicate, better known as 'water glass', and often used in preserving eggs, has wide applications in both industry and the home. Today it is manufactured by fusing natron (Na_2CO_3) and silica (SiO_2 at high temperatures.

Davidovits and Morris (1988) indicate that the substance was first produced during the fourth millennium, some 5600 years ago. However, the nature of the quartz, or silica, used in its production needs to be carefully selected for optimum processing, hydrous silicates being preferred. Chert is such a type of silica, and this rock type abounds in the hills around the Dead Sea.. The manufacture of sodium silicate is expressed by the formula:

$$2NaOH + SiO_2 \rightarrow Na_2SiO_3 + H_2O$$

The fact the Egyptians could manufacture glass strongly suggests that they might also have been able to produce sodium silicate. Sodium silicate reacts with materials containing calcium to produce calcium silicate ($CaSiO_3$), and whereas the former is water soluble the latter is insoluble. This is the reason why water glass is so effective in preserving eggs – the sodium silicate reacts with the calcium carbonate of the eggshell, which being porous leaves an egg open to bacterial infection. By immersing eggs in water glass the following reaction is promoted:

$$CaCO_3 + Na_2SiO_3 \rightarrow CaSiO_3 + Na_2CO_3$$

The calcium silicate is deposited within the pores of the eggshell thereby inhibiting ingress of adulterating matter. A similar reaction occurs when the surface of imperfect and porous concrete is treated with a silicate solution and, thereby, rendered impermeable.

Another important use of sodium silicate in modern construction is in the grouting of foundation soils, where injection of the silicate takes place in conjunction with calcium chloride. Once again calcium silicate is produced rendering the soil less porous. The process is represented by the following equation:

$$CaCl_2 + Na_2SiO_3 \rightarrow CaSiO_3 + 2NaCl$$

It is possible, therefore, but by no means certain, that techniques used by the Egyptians in the construction of the pyramids also may have utilised silicates. Desert sand is plentiful. By mixing the sand with calcium chloride to obtain a uniform mix, hoisting it into place, and saturating the forms with a solution of sodium silicate, it is perfectly feasible to have produced an acceptable artificial stone. Such a possibility would support Davidovits's general thesis.

Beer and Wine

There is a legend that Dionysos, the Greek god of wine and barley, fled Mesopotamia in disgust because the inhabitants were so addicted to the consumption of alcohol in its varied forms. According to Forbes (1964/72 vol II) both beer and wine were produced across the vast region from Assyria to Egypt, old Sumerian texts making reference to eight types of barley beer, eight types of emmer beer, and a variety of other beers.

Vineyards flourished throughout Northern Mesopotamia, with Harran, a centre of wine production boasting vineyards containing up to 29,000 vines. Vineyards in Egypt first appeared within the Delta region extending as far as Memphis, but later spread southwards up the Nile valley. The importance of viniculture to daily Egyptian life is reflected by the existence of a tomb of the First Dynasty (3100-2900 BC) where the body was placed upon a layer of grapes.

It will never be known to what extent the Sodomites may have practised viniculture or brewing, but in light of their revels outlined in the Book of Jasher it would not be surprising if alcohol played a major role in the general licentiousness.

Commentary

Though the contents of this chapter are admittedly highly speculative, it is hoped that it serves the purpose to demonstrate the relative simplicity of the chemical engineering involved in the enhancement of the raw materials readily available to the Sodomites. Therefore, there is good reason to believe that, with the aid of Egyptian alchemists, they could have developed an extensive industrial complex, primitive though such an industry might appear to modern eyes. The chemical works located at the southern end of the Dead Sea today are testimony to the richness of the resources in the area. The long caravans toiling up the Western Ascent out of the Dead Sea, and winding across the arid wilderness *en route* to Egypt, may have carried bitumen as a mainstay of trade, but sulphur, natron, carnallite, and silicate, may have been useful commodities also carried thither.

It is hoped that at some future date archaeologists will discover further evidence of a strong interlinking commerce between Canaan and Egypt, based upon trade in the basic commodities discussed herein, all of which would have been important to the economy of Egypt. Such evidence would confirm the suggestion that the Sodomites were among the world's first chemical engineers.

As Ecclesiastes (1:9) wrote – 'The thing that hath been, it is that which shall be; and that which is done is that which shall be done: and there is no new thing under the sun.'

7. Earthquakes and Landslides

Earthquake Magnitude and Intensity

Two common scales for evaluating and comparing earthquakes are (a) The Richter Magnitude Scale, and (b) The Modified Mercalli Scale of Felt Intensity. The terms 'magnitude' and 'intensity' have specific meanings, and to avoid confusion the two scales are fully explained in Appendix II.

The Dead Sea Graben and Seismicity

One of the best known geographic features of Earth's crust is the Great Rift Valley of East Africa, which is considered to have developed during the Miocene approximately twenty million years ago. The result is a continuous 'trench', known as a graben, approximately 6000 kilometres (3700 miles) long extending from the mountains of Anatolia to central Mozambique. The northern end, known as the Dead Sea Rift, includes the Sea of Galilee, the Jordan River, the Dead Sea, the Wadi Arava, and the Gulf of Aqaba. A significant number of earthquakes are associated with this fault, which forms the boundary between two major tectonic plates, i.e. the Palestinian Plate to the west, and the Arabian Plate to the east (see Figure 39).

Relative displacement between the two plates due to strike-slip faulting, is considered to have commenced during the Late Cretaceous, the total displacement along strike now amounting to approximately 100 kilometres (about 60 miles), the Arabian Block having moved north relative to the Palestinian Block. The forces creating movement between the two plates, however, are not only shear but tensile as well. Thus, there is a force component tending to pull apart the two tectonic blocks, the net effect being a tendency to widen the Dead Sea valley.

East of the Dead Sea Rift are the mountains of Edom, to the west are the Judean Hills. The former attain significantly greater heights than the latter, rising to almost 1600 metres. Whereas the mountains of Edom are older crystalline and metamorphic rocks, which at sunset often glow red (the Hebrew for 'red' being 'adom'), the Judean Hills are mainly composed of younger limestones. This disparity in geology between the two sides of the Dead Sea is a consequence of this strike-slip faulting.

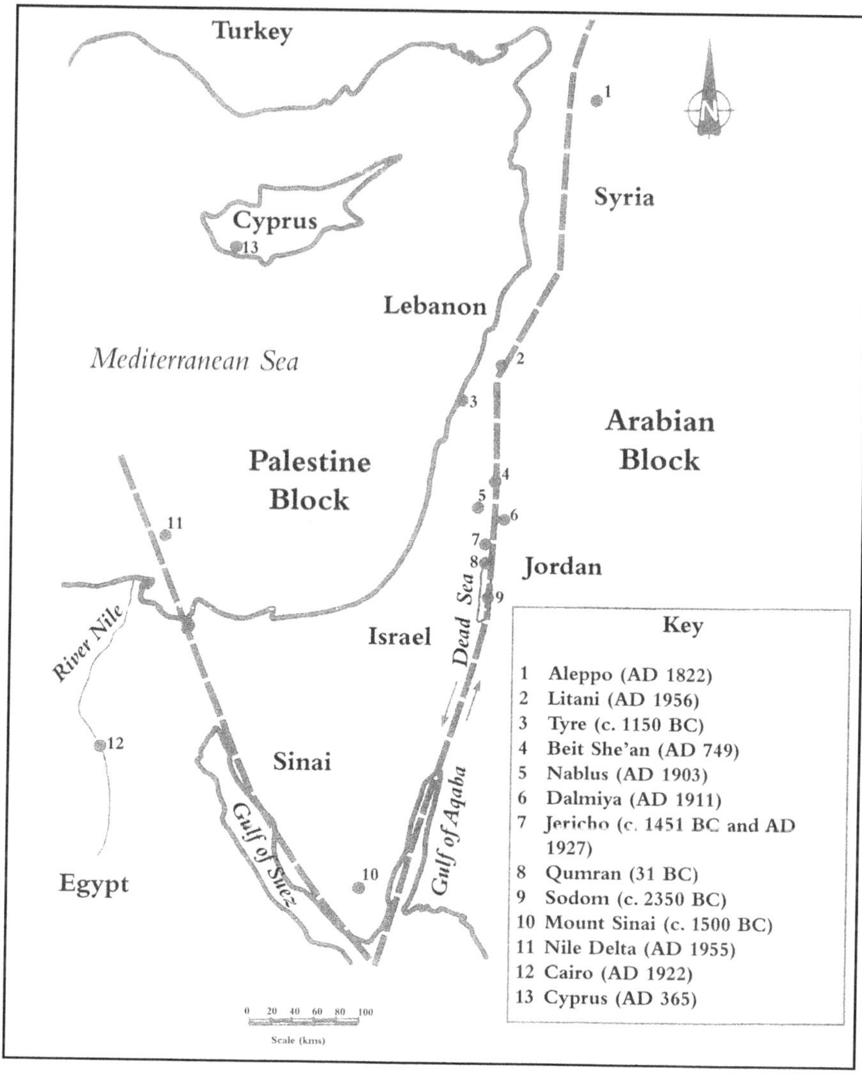

39. Regional seismic map with a selection of historic earthquakes.

The enormous shear displacements associated with the north-south sinistral movement of the two tectonic blocks has resulted in the formation of a wide fault zone, known geologically as the Dead Sea Transform. Within this zone there is a great deal of transverse and subsidiary faulting, frequently arranged *en echelon*, which has resulted in a complex fault pattern beneath the Dead Sea. This shear zone is bounded on east and west by major faults known respectively as the East and West Border Faults. The former is more or less coincident with the eastern shoreline, whereas the latter is set back somewhat. As a result the eastern side of the valley rises more steeply from the edge of the Dead Sea than its counterpart to the west, and this has had an influence on trade routes.

Transverse faulting occurs at high angles of incidence within the highly fractured shear zone demarcated by the East and West Border Faults. As a consequence of the complexity of forces there is considerable discontinuity between adjacent blocks of ground following the north-south trend of the valley. The Lisan Peninsula, strategically located between the North and South Basins, is one such block (known as a *horst*) which is the result of uplift.

Biblical Earthquakes

The first reference to an earthquake in the Bible, is that implied by the verse in Exodus (19:18) which reads, 'And Mount Sinai was altogether on a smoke, because the Lord descended upon it in fire: and the smoke thereof ascended as the smoke of a furnace, and the whole mount quaked greatly.' While it is difficult to ascribe intensity to this event on the Richter Scale, the description would appear to warrant an intensity of at least VIII on the Modified Mercalli Scale.

It is of interest to ponder the events relating to the fall of the walls of Jericho. Joshua (4:13) gives the size of the Israelite army as 'about forty thousand prepared for war'; while verse 6:20 recounts 'So the people shouted when the priests blew with the trumpets: and it came to pass, when the people heard the sound of the trumpet, and the people shouted with a great shout, that the wall fell down flat.'

It is doubtful whether the Israelites, even inspired by emotional fervour, could have generated sufficient seismic energy necessary to flatten the walls of Jericho by shouting alone. Since the walls fell down flat, the event warrants an intensity of at least X on the Modified Mercalli Scale. Forty thousand Hebrew warriors, each weighing a hundred kilograms clad with armour, and simultaneously leaping a half metre into the air, would have generated a shock loading no more than that released by a moderate seismic event rating only a magnitude (M) of about 3 on the Richter Scale. However, it must be noted that Jericho lies within a kilometre of the Western Border Fault of the Dead Sea graben. The proximity of this major fault, and its swarm of associated faults, suggests the Israelites besieging Jericho might have had some seismic assistance. The juxtaposition of the two could well have constituted the miracle ascribed to the collapse of the walls, any seismic shaking having gone unnoticed in the general euphoria.

Zechariah (c. 500 BC) writes in verse 14:14, 'and the Mount of Olives shall cleave in the midst thereof toward the east and toward the west, and there shall be a very great valley; and half the mountain shall remove toward the north, and half of it toward the south.' If ever there was a description of a classic wrench fault, then this is it! Amos (c. 800 BC) also writes in verse 1:1 'the words of Amos . . . in the days of Uzziah, King of Judah . . . two years before the earthquake'. These two separate accounts, written three centuries apart, both suggest high value seismic events.

Earthquakes in the Holy Land

For a comprehensive listing of earthquakes in the Holy Land, reference may be made to Ben-Menahem (1979). Also, Nur and MacAskill (1990) have discussed many of these seismic events in a fascinating video titled *The Walls Came Tumbling Down*. The various listings, however, are unlikely to be exhaustive, with many earthquakes remaining unrecorded, or recorded only through folklore, with little documentation, if any, from which to ascribe either magnitude or intensity. Evidence of their occurrence, however, may occasionally be detected in the geologic or archaeologic record, and these represent rich sources for research into past seismicity. Table E[1] lists some of the notable historic events indicative of the seismicity of the region, whereas Figure 40 shows epicentres of earthquakes within the Dead Sea in the period 1907 to 1993.

Location	*Year*	*Intensity*	*Magnitude*	*Remarks*
Sodom	c. 2350 BC	>IX	>6.0 (est)	Total destruction (Genesis 19)
Crete	c. 1700 BC	unknown	unknown	Palace of Knossos destroyed
Mount Sinai	c. 1500 BC	>VIII	unknown	Mountain shaken (Exodus 19)
Jericho	c.1450 BC	>X	unknown	City walls tumbled (Joshua 4-6)
Tyre, Lebanon	c.1150 BC	IX-X	unknown	Temple of Melkart destroyed
Holy Land	c. 800 BC	>X	unknown	Widespread terror (Amos 1)
	c. 500 BC	XI-XII	unknown	Mountains split (Zechariah 14)
Dead Sea	64 BC	unknown	>5.5	Kan-Tor *et al.* (2001)
Qumran, Israel	31 BC	>X	>5.5	Total destruction (Josephus)
Dead Sea	33 BC	unknown	>5.5	Kan-Tor *et al.* (2001)
Antioch, Syria	AD 115	unknown	unknown	Closson *et al.* (2004)
Crete and Cyprus	AD 365	>IX	8.0 (est)	Kelly (2004)
Antioch, Syria	AD 525	unknown	unknown	Closson *et al* (2004)
Beit She'an, Israel	AD 749	>X	unknown	Destruction of the Roman city
Pella, Jordan	AD 749	unknown	unknown	Closson *et al.* (2004)
Tripoli, Lebanon	AD 1063	unknown	unknown	Closson *et al.* (2004)
Ramla, Israel	AD 1068	unknown	unknown	Closson *et al.* (2004) – also suffered earthquake damage 715, 1033, 1546 and 1927 *(Table E continued on next page)*

1. Note: After Harris and Beardow, 1995, Kan-Tor *et al.*, 2001, and Closson *et al.*, 2004 (1) Intensity in Modified Mercalli Scale units (approx) (2) Magnitude in Richter Scale units (3) Dates of biblical events are best estimates.

Location	*Year*	*Intensity*	*Magnitude*	*Remarks*
Holy Land	AD 1170	>X	unknown	Towns and castles destroyed
Baalbeck, Lebanon	AD 1202	unknown	unknown	Closson *et al.* (2004)
Dead Sea	AD 1212	unknown	>5.5	Kan-Tor *et al.* (2001)
	AD 1293	unknown	>5.5	Kan-Tor *et al.* (2001)
Baalbeck, Lebanon	AD 1759	unknown	unknown	Closson *et al.* (2004)
Aleppo, Syria	AD 1822	unknown	unknown	22,000 deaths
Dead Sea	AD 1834	unknown	>5.5	Kan-Tor *et al.* (2001)
Nablus, Israel	AD 1903	unknown	5.7	
Damiya, Jordan	AD 1911	unknown	6.0	
Cairo, Egypt	AD 1922	X	5.9	
Crete	AD 1922	unknown	6.5	
Jericho	AD 1927	>X	6.2	Widespread destruction
Nile Delta	AD 1955	unknown	6.3	
Litani, Lebanon	AD 1956	unknown	5.8	
Dead Sea	AD 1979	unknown	5.1	SE corner of Dead Sea
Nuweiba, Aqaba	AD 1995	unknown	unknown	Closson *et al* (2004)

Table E – A selection of regional earthquakes including Biblical events

It may be noted that the greatest concentrations of seismic activity are within the Gulf of Aqaba and in proximity to the Dead Sea. These are obviously directly associated with the Rift Valley system. Neev and Emery (1995) comment upon the earthquake that likely led to the destruction of Sodom in the following terms:

> Three factors control most effects of an earthquake at a given site: distance of the epicentre from the site, frequency-magnitude relations, and physical characteristics of the subsurface such as rock density, mineralogy, matrix-to-coarse-detritus ratio, compaction, and liquefaction-slide processes. Ben-Menahem *et al.* (1977) and Ben-Menahem (1981) considered that magnitude of the Sodom and Gomorrah earthquake was more than 7.0 on the Richter scale somewhere along the Dead Sea rift. It may have been the strongest earthquake in this region during the 6000 years of historical time. Biblical description of this event, the sole source of eyewitness information, was only an oral tradition for several thousand years.

Earthquakes resulting from tectonic displacement within the Dead Sea Rift are numerous. However, the great majority are weak, of low magnitude, and scarcely felt. Approximately, only a quarter of those recorded liberate sufficient energy to be of much concern, and few of those powerful enough to result in damage. Over the last two thousand years it is estimated only about seventeen events were sufficiently damaging to be rated ruinous.

In recent decades a great deal of research has been carried out by seismologists, and these have concluded that (a) most of the earthquakes beneath the Dead Sea (about 60 percent) originate at depths between 20-32 kilometres, most of the remainder being shallower than 20 kilometres; (b) that the return period of significant earthquakes, possessing a magnitude (M) of 5.8 or more, is of the order of 160 years (Shapira, 1997); (c) the latest earthquakes of any great severity were those which were associated with the Jericho event of 1927, which had a magnitude of M=6.2 (Turcotte and Arieh, 1988), and on the Carmel Fault, located 16 kilometres from Megiddo, in 1984 (Nur and Ron, 1997).

Furthermore, Nur and Ron argue that three major earthquakes struck Canaan between 1550 and 1000 BC from the evidence of crushed human remains and shattered jars discovered under collapsed walls. Such a return period for major earthquakes is in general agreement with Shapira (1997).

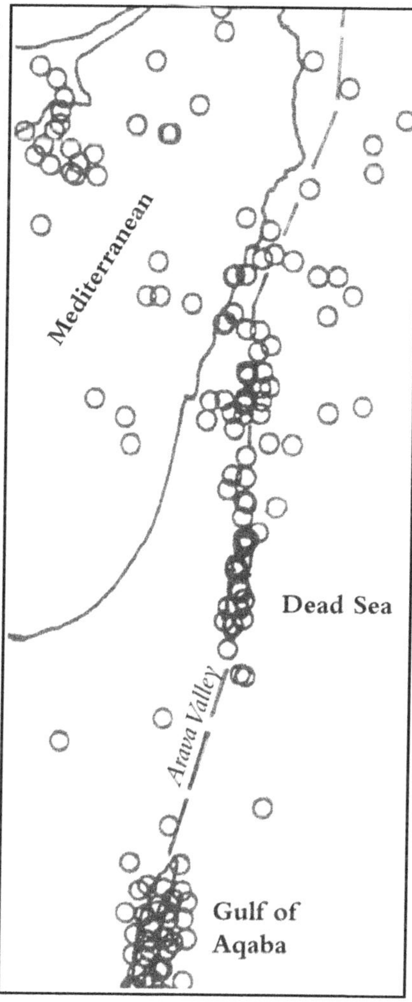

40. Epicentres of many earthquakes within the Dead Sea valley (M≥4) between 1907-93, compiled by the Seismological Division of the Israel Institute of Petroleum Research and Geophysics (after Neev and Emery, 1995).

Enzel *et al.* (2000) in their paleoseismic studies at Nahal Darga, in the northeast of the Dead Sea valley, concluded that during the last century only five earthquakes had occurred with a magnitude of M>5, of which the greatest was the Jericho event of 1927 of magnitude M=6.2. Their studies were based on observations of slumping characteristics of sediments exposed in the eroded sides of wadis, which could be related to earthquake activity. The sandy layers were of a gradation which, when saturated, were prone to liquefaction triggered by seismic tremors. The nature of the exposures were such that a number of faults could also be identified which hinted of significant seismic movement. These faults possessed vertical components ranging from a few centimetres to two metres, with larger displacements in older units. From all the evidence, including radiocarbon dating, it was concluded that the timing of some of the older faulting suggested dates of between 9500-7000 years BP, with a return period of about six hundred years for earthquakes with a magnitude of M=5.5.

Despite the above, Marco *et al.* (1996), by studying exposures of Lake Lisan sediments provided evidence to suggest that the mean return period of earthquakes with similar magnitudes was somewhat longer being of the order of 1600 years, but clustering took place at intervals of approximately 10,000 years, with relatively quiescent periods between more active periods of seismicity.

Ken-Tor *et al.* (2001), in their turn, studied the evidence for past earthquakes in lacustrine sediments accumulated beneath the Ze'elim Plain, on the western side of the Dead Sea. They discovered eight layers showing deformed structures which could be associated with earthquake activity. Such geologic abnormalities are known as seismites. They identified organic remains embedded within the affected layers, and determined their ages by radiocarbon dating. Though the resultant record was incomplete, the findings correlated well and supplemented earlier work. One of the more important conclusions was the indication of a return period of earthquakes possessing a magnitude M=5.5 in the order of every 100-300 years. Secondly, that the earthquakes were a consequence of slip-strike movement along the various faults beneath the Dead Sea basin.

Hartel *et al.* (2003), in connection with the AD 749 earthquake at Beit She'an, confirmed the findings of previous researchers, and reported that results from paleoseismic studies were in good accord with documented historical earthquakes. They also concluded that the geology of the Jordan River valley was capable of generating earthquakes M=7.

Closson *et al.* (2004) focussed their attentions on the seismicity of the Dead Sea region with a view to identifying the geologic hazards resulting from tectonic movements, particularly as these might affect industrial development on the Jordanian shore. Two years earlier the Arab Potash Company had experienced some instability problems in connection with the northward expansion of their solar evaporation ponds on the west side of the Lisan Peninsula. This expansion halted on March 22, 2000, when a 2.3 kilometre long section of dyke failed.

One of the more interesting historic earthquakes in the Near East region, which resulted in a massive tsunami, had its epicentre offshore Crete. This underwater event, which occurred in AD 365, was described by the Roman historian Ammianus Marcellinus (c. 325-391), see Kelly (2004). It destroyed much of the port of Alexandria, boats being plucked out of the ocean and flung two miles inland.

Another event of antiquity having considerable significance to the overall thesis of this book, is the total loss of the ancient Achaean city of Helike (now Eliki) in Greece, which occurred 373/2 BC. Seed (1967) has attributed the loss of the city to soil liquefaction initiated by earthquake. The event was repeated in 1861 with less disastrous effect.

More Recent Earthquakes

Though earthquakes associated with the Great Rift Valley in general, and the Dead Sea graben in particular, can be exceedingly destructive, the largest and most destructive earthquakes have been experienced elsewhere. The most powerful event on record is that of Valdivia, Chile (May, 22, 1960) which registered a magnitude M=9.5, killing nearly 6000 people, and resulted in a tsunami which battered the shores of Easter Island. An Indonesian earthquake off the coast of Sumatra (December 26, 2004) was of similar magnitude (M=9.2), but far more destructive, taking a toll of about 250,000 people the resulting tsunami wreaking havoc around the Indian Ocean. The Port-au-Prince, Haiti, earthquake (January 12, 2010) had a magnitude of M=7.0 and ruined this desperately poor country. This was immediately followed by the Concepcion, Chile, earthquake (February 27, 2010) estimated as having a magnitude M=8.8.

In recent times an event of far-reaching impact (M=9.0) was that which occurred at Tohoku, Japan, on the 11 March 2011. This resulted in a massive tsunami which drove inland severely damaging the Fukushima Daiichi nuclear plant, becoming the world's worst disaster in the history of nuclear power generation. The focus of the earthquake was about 70 kilometres off the east coast, at a depth of 30 kilometres. Close to 20,000 lives were lost as a consequence when the hitherto tranquil region was transformed into ravaged wasteland. The earthquake was preceded two days earlier by a foreshock (M=7.3) and followed by numerous high magnitude aftershocks, several having magnitudes exceeding M=6, and three with magnitudes above M=7. Over 900 aftershocks were recorded in the nine months following the event. Much of the debris from the tsunami was swept out to sea and, at the time of writing, much of this is washing up on the Pacific coasts of Canada and the United States.

Table F lists some of the world's most lethal earthquakes experienced during the past century. The Indonesian event is not included on this list, as the majority of the deaths resulted from the ensuing tsunami.

Year	*Location*	*Magnitude*	*Death Toll*
2011	Tohoku, Japan	9.0	c. 20,000
2010	Port au Prince, Haiti	7.0	227,750
1995	Kobe, Japan	7.2	6348
1990	Western Iran	7.7	c. 50,000
1976	Tangshen, China	8.0	290,000
1970	Chimbole, Peru	7.7	c. 75,000
1937	Erzincan, Turkey	8.0	36,740
1936	Quetta, Pakistan	7.5	35,000
1927	Xining, China	8.3	200,000
1923	Tokyo-Yokohama, Japan	8.3	142,800
1920	Ganzu, China	8.6	235,000

Table F – A Selection of major earthquakes with the highest death tolls

Submarine Landslides

Immense volumes of submarine materials may be loosened by seismic events. Deltas are very prone to slides as they are regions of sediment accumulation and, therefore, weak and unstable by virtue of their unconsolidated nature. Even when not initiated by seismic forces submarine landslides can be triggered by offshore currents down-cutting into recently deposited sediments, thereby causing small slides which can enlarge progressively to encompass significant volumes of material.

Numerous submarine slides occur annually, especially at locations where the flow of major rivers, e.g. the Mississippi, St. Lawrence, Ganges, and Amazon, which accumulate vast quantities of sediment in the underwater extensions of their delta systems, cross their respective continental shelves. Many of these slides take place relatively slowly, often assuming characteristics of turbidity flows and, therefore, of little practical significance except, perhaps, to oil-drilling platforms operating in the affected areas.

Certain regions of the world are exceedingly vulnerable to submarine landslides, and other associated sea floor displacements, especially Japan where approximately a quarter of the world's earthquakes are experienced with magnitudes greater than M=6. Following the 1923 earthquake that devastated Tokyo and Yokohama, the sea floor in Sagami Bay dropped as much as 300 to 400 metres in some places, other parts being raised. Other regions, notably those also associated with 'the ring of fire' encompassing the Pacific Ocean, are similarly prone to offshore seismic activity.

Sediments Prone to Liquefaction

Sedimentary soils arise from a cycle of erosion, transportation, and sedimentation. The subsequent strength of the resulting soil may bear no relation to the rock type or soil from which it originated, a variety of environmental factors influencing its final composition. Soils are generally categorised as sand, silt or clay, or mixtures of these, depending on their gradations – these three categories being defined by strict gradation limits. The degree of compactness of a soil, however, is reflected by its void ratio, or porosity, the higher these values the less compact the soil. Much depends on the nature of the sedimentation process, and its subsequent history, as to whether a soil remains loose or dense, thereby controlling its structural strength to a major degree. Many soils become consolidated by subsequent layers of deposited sediment, or overburden, the weight of which compacts the soil, thereby reducing its void ratio, or porosity, and increasing its density and enhancing its strength.

Some soils, notably clays, possess a significantly higher strength when they are 'undisturbed', i.e. when their inter-particle contacts are not destroyed. When carried out in the laboratory this destructive process of the soil fabric, or matrix, is termed 'remoulding'. Figure 41 vividly demonstrates the loss of strength in a sample of Leda clay (obtained from the Ottawa valley, Canada) after being remoulded. The ratio of undisturbed strength to remoulded strength is termed the 'sensitivity' of that particular soil, and for the Leda clay shown this is about 30. This is, however, an extreme example of a highly sensitive clay.

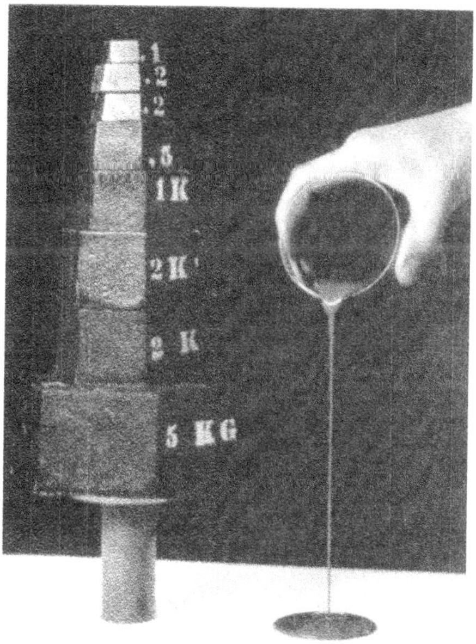

41. Demonstration of the intact and remoulded strengths of a highly sensitive clay (Crawford, 1963).

Seismic forces, through the vigour of their shaking, can reduce soil strength significantly and suddenly, and this is often a cause of extensive loss of ground, or lateral spreading, when what was once seemingly solid ground is liquified. The higher the sensitivity of a soil the more prone is this loss of strength. When the strength is reduced to such a low value that it begins to flow like a liquid the phenomenon is termed 'liquefaction'. Liquefaction can only take place in loose soils susceptible to contraction of the soil matrix, and which are highly saturated. The liquefied soil will begin to flow like a viscous liquid after its strength is reduced to the extent that it cannot resist the forces involved.

A simple demonstration of liquefaction has been experienced by everyone walking along a beach close to a receding tidal edge. Few have resisted the childish pleasure of jumping up and down on fine sand saturated with water, for the simple fascination of watching their feet gradually sink into the liquefied sand, and watch the water ooze out. On a much larger scale this everyday phenomenon has been the cause of great disaster in the form of mudslides, many triggered by seismic forces, which have devastated numerous communities.

One of the first observations of this causal effect was that noted by G.M. Dawson (1899), who described such a slide at St. Thuribe, Quebec, in the following terms:

> It appears probable that in this particular instance the silty-clay, surcharged with water, stood in a condition of unstable equilibrium, retaining its solidity merely by virtue of its unbroken molecular texture, and at the moment in which it became subject to internal movement this texture gave way and it lapsed into a nearly liquid mass, the particles rearranging themselves with some freedom in the water previously locked up in its pores.

It is now known that the sensitivity of a silt or clay soil depends to a large extent upon whether the soil has been deposited in fresh or salt-laden water, the latter having the effect of creating a flocculated internal arrangement of the individual soil particles due to the electro-chemical forces involved. Such soils may be visualised as possessing a 'card-house structure'. Invariably they are identified with higher sensitivities and are subject to collapse, or liquefaction, when subjected to shock loadings, such as those caused by earth tremors or seismic forces. Saturated fine-grained sands also possess a similar ability to liquefy, but as the sand particle size increases so the potential for liquefaction lessens.

Many disastrous landslides of major consequence have had their origin in the phenomenon of liquefaction. There is no region of the world more prone to this phenomenon than any other. However, areas of the Northern Hemisphere underlain by clays deposited in salt-water environments, as a consequence of glacial melting during the post-glacial period, have had their fair share of incidents.

Reference was made in Chapter 2 to the Sedom Clay, which has been shown to underlie recent sediments bordering the Dead Sea at shallow depths, and to extend to depths in excess of 40 metres. Because of the highly saline nature of the waters in which the Sedom Clay was deposited it is not surprising that the deposit possesses a degree of sensitivity making it somewhat prone to liquefaction. This characteristic is believed to have played some role in the postulated liquefaction event of c. 2350 BC resulting in the loss of Sodom, Gomorrah and the other 'cities of the plain'.

Liquefaction Events of the Past

Seed (1967) draws attention to the wholesale destruction of Helike, Greece, in 373/2 BC, attributing this to soil liquefaction by earthquake, an event dramatically repeated two thousand years later at the same locale in 1861. It is worthwhile giving due attention to this particular catastrophe, as it not only mirrors that of the Sodom event postulated here, but because many of our cities are located on liquefiable soils and, possibly, subject to a similar fate. Seed's hypothesis is based upon the description of Marinatos (1960) who writes:

> In the year 373/2 *BC*, during a disastrous winter night, a strange thing happened in central Greece. Helice, a great and prosperous town on the north coast of the Peloponnesus, was engulfed by the waves after being levelled by a great earthquake. Not a single soul survived. . . . The next day two thousand men hastened to the spot to bury the dead, but they found none, for the people of Helice had been buried under the ruins and subsequently carried to the bottom of the sea, where they now lie.

Seed suggests this was one of the earliest known cases of a major landslide attributable to soil liquefaction by earthquake. He writes:

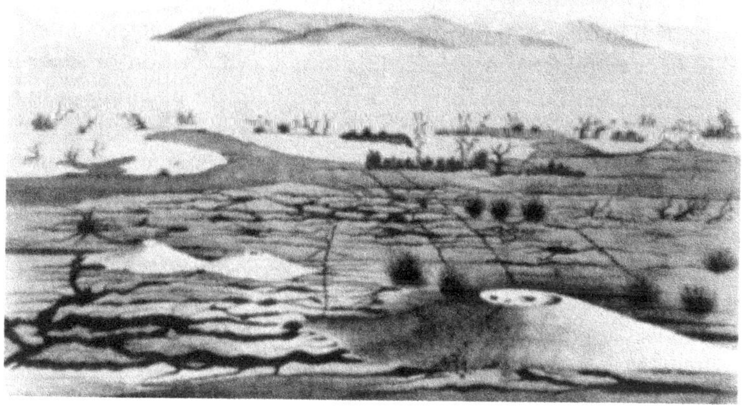

42. Sand boils at Helike, Greece, following the liquefaction event of 1861 (Schmidt, 1875).

Helike was located on deltaic deposits of alluvial sand between the mouths of the Selinus and Cerynites rivers and about a mile and a half from the coast. The city is now completely covered and no trace of it exists, neither on the ground surface nor on the bottom of the sea.

The events leading to the disappearance of Helike and its inhabitants are not immediately clear. A general subsidence of the land area during the earthquake undoubtedly occurred and this alone could have led to flooding of the city. However both Schmidt (1875) and Marinatos (1960), who made detailed studies of the event, concluded that in addition to destruction of buildings by the ground shaking and flooding due to land subsidence, the ground slipped towards the sea possibly as much as half a mile. Marinatos notes that ordinarily it would be expected that building destruction and flooding would lead to some of the dead floating to the surface where they would have been picked up by the Achaeans for burial. It seems reasonable to conclude that only the entrapment of the inhabitants in collapsed buildings, and temporarily liquefied and flowing soils could have led to the recorded facts that no one survived and no dead were found.

In the words of Marinatos:

> the phenomenon was repeated, in exactly the same place though to a lesser degree, during the earthquake of December 26, 1861. . . . Again the soil slipped to the northeast (toward the sea) in the following way: a crack about eight miles long and six feet wide appeared in the earth along the foot of the mountain. A strip of plain 325 ft to 425 ft wide disappeared slowly under the sea along the eight-mile length, while the remaining part of the plain sank about six feet and showed many minor cracks and small chasms.

Sketches prepared by Schmidt, during his investigation of the results of the 1861 earthquake, point to the part played by soil liquefaction. One of these sketches reproduced in Figure 42 shows the formation of sand boils at ground surface. Sand boils and sand craters are frequent phenomena following liquefaction events, and the existence of either is considered an indication of liquefaction of the underlying soils having taken place. Figures 43 and 44 are examples of the considerable damage to surface structures and road pavements that may result from liquefaction leading to massive ground displacement during such events.

Many significant liquefaction events may be noted in the historical record. Seed (1967) lists another thirty-six examples of earthquake induced liquefaction that occurred between 1755 and 1966. Many of these resulted in massive loss of life, especially that of Lisbon, Portugal, 1755, where the destruction was so widespread from the resultant tsunami waves, that it is popularly believed to have spurred study into earthquakes, leading to what is now known as seismology.

One of the most dramatic descriptions of ground liquefaction is that supplied by Close and McCormick (1922) from the Kansu, China, earthquake of 1920. They write:

> Of that most remarkable series of seismic disturbances which occurred through-out the world in November and December, 1920, the most phenomenal was undoubtedly the great Kansu earthquake of the late evening of December 16. . . . Landslides that eddied like waterfalls, crevasses that swallowed houses and camel trains, and villages were swept away under a rising sea of loose earth, were a few of the subsidiary occurrences that made the earthquake in Kansu one of the most appalling catastrophes in history.
>
> The area of greatest destruction, 100 miles by 300 miles in extent, contains ten large cities besides numerous villages. It is in the heart of loess country . . . where the loose earth cascaded down the valleys and buried every object in its path.
>
> It is in the loess area that the immense slides out of the terraced hills occurred, burying or carrying away villages . . . damming stream-beds and turning valleys into lakes, and accomplishing those hardly believable freaks which the natives name 'the footsteps of the gods.' The loss of nearly two hundred thousand lives and the total loss of hundreds of towns and cities calls for reconstruction work on a staggering scale.

In one valley Close and McCormick describe a singular occurrence of survival amidst all this disaster:

> The only survivors in this valley were saved as if by a miracle – a husband man and his two sons, whose farmstead, instead of being buried, was caught upon the back of one of the slides, carried half a mile down the valley to

43. Loss of ground by liquefaction following the Anchorage, Alaska, earthquake, 1964 (USGS).

44. Foundation failure from liquefaction, Adapazari, Turkey, earthquake 1999 (USGS).

where it was diverted by two streams of earth coming from other directions and, as a result of the two forces, was pushed another quarter of a mile up a small draw.

Not until day dawned and they crawled out to find neighbouring villages obliterated, farm lands carried away or buried, streams blocked and hills of earth towering above their compounds did they apprehend that the 'hills had walked'.

Total liquefaction of a soil is unnecessary in order to produce a liquefaction slide. Provided the earth contains a stratum, or layer, or even a thin lamina of material that becomes liquefied, significant sliding can ensue, the overlying soil floating like blocks upon the liquefied soil, as may be visualised from the descriptions of Close and McCormick. Such was the case in several of the landslides experienced in Alaska in 1964 when an 8.3 magnitude earthquake struck. The port of Valdez was mostly affected, being built on delta deposits consisting of silt, fine sand and gravel, with the silt and fine sand distributed as thin beds and stringers within the coarser sediments. Approximately 150 million tons of soil was displaced seawards. The slide extended about five hundred feet inland, and destroyed harbour facilities and near-shore installations (see Figure 43). In some places the depth of soil involved in the slide extended two hundred feet below surface as it moved several thousand feet outwards into the bay.

Basic Properties of the Sedom Clay

Adler and Harris (1983) showed the Sedom Clay to be a young micritic limestone. Associated with the calcite, which comprises more than 90 percent of all samples, are detrital quartz, ferro-magnesian particulates, and salt. Additional minor constituents include gypsum, clay minerals, and other carbonates. Occasional laminae of fine sand may be encountered throughout the stratum. The micrite would not have been precipitated in one continuous sequence, but intermittently over the several thousand years between when the water level of Lake Lisan, the predecessor of the present Dead Sea, fell from its highest to its lowest levels.

Because of its origin as a chemical precipitate, the micrite may be expected to be extensively distributed draped over previously deposited underlying sediments, and local variations in composition and difference in physical properties are inevitable. Actual thicknesses of the micrite are likely to be highly variable, and may even be absent where it has been eroded by wadis discharging into open water, or where they have carved out submarine channels within the micrite which, subsequently, have become filled with alluvium.

The recent micrites, lying higher in the soil column, possess high void ratios and high moisture contents rendering them susceptible to liquefaction. The variation of these two parameters from surface to a depth of ten metres is indicated on Figure 45. With the exception of the single uppermost value, which may indicate some crustal desiccation, both parameters trend lower with depth. The highest values, at a depth of five metres, reflect a void ratio of 1.25 and moisture content of 42 percent respectively. Other geotechnical parameters are summarised on Table G. A consideration of particle size distribution indicates the total silt and clay content to average 85 percent, but in view of the fact that the calcareous particulates possess a rhombohedral shape the Sedom Clay, despite its name, is lacking in most of the characteristics possessed by clays as commonly accepted.

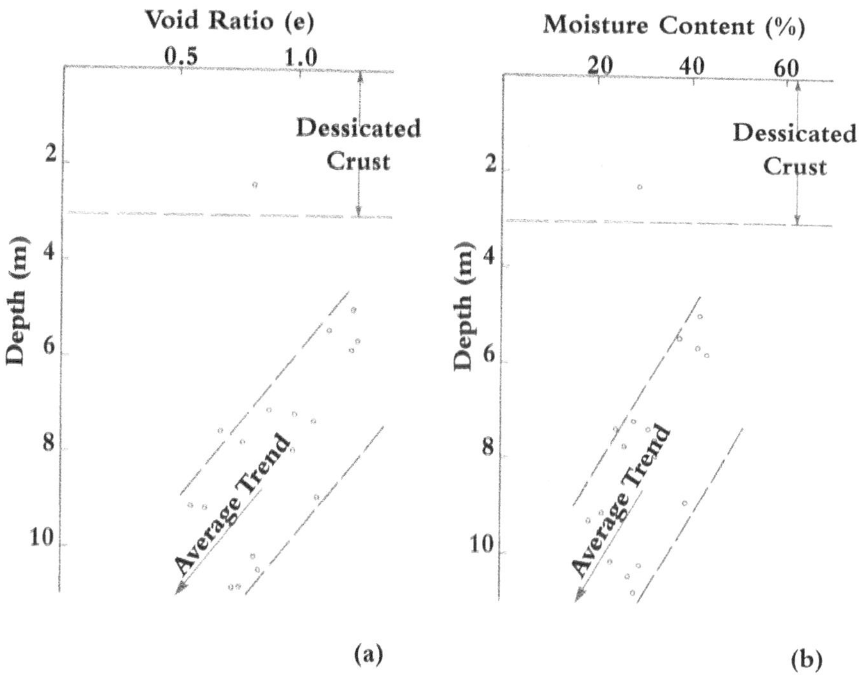

45. Variation with depth of (a) void ratio, and (b) moisture content of the Sedom Clay.

Liquefaction Potential of the Sedom Clay

Wang (1979) indicated certain types of clayey materials to be vulnerable to severe loss of strength when subject to seismic forces. He suggested this vulnerability is controlled by the following characteristics:

Material finer than 0.005 mm ≤ 15 percent
Liquid Limit ≤ 35 percent
Water content ≥ 0.9 x liquid limit (%)

The in situ properties of the Sedom Clay do not fully meet these requirements, especially with respect to the percentage of fines, but then the Sedom Clay is not a true clay. Otherwise, the low average liquid limit and generally high water content conform well with Wang's criteria. The presence of sand laminae within the stratum, as noted early, would enhance the susceptibility of the deposit to liquefaction under seismic loading. Values of vane shear strength (C_u) for the Sedom Clay, plotted against depth from ground surface, as obtained from a borehole put down on the property of Dead Sea Works at S'dom, Israel, are shown on Figure 46. On the same figure are plotted computed values of cyclic shear stress (τ_{av}) for varying values of seismic ground acceleration. It may be observed that, for accelerations of between 0.6g to 0.8g, the intact strength of the soil begins to be exceeded. Since the sensitivity of the soil ranges between 1.9 and 4.0, a rapid and progressive loss of strength could be expected as a

Soil Property	*Number of Tests*	*Range*	*Average Value*
Gradation			
Sand	9	1-37%	14%
Silt	9	28-55%	42%
Clay	9	22-71%	56%
Silt and Clay	9	63-99%	86%
Moisture Content	29	19.2-43.6%	32%
Density			
Bulk	30	1759-2116 kg/m3	1931 kg/m3
Dry	30	1242-1775 kg/m3	1475 kg/m3
Specific Gravity	18	2.76-2.89	2.80
Atterberg Limits			
Liquid	18	18-43	32
Plastic	18	15-25	20
Plasticity index	18	3-21	12
Undrained Strength			
Undisturbed (Vane)	8	28-104 kPa	69 kPa
Remoulded (Vane)	8	12-54 kPa	26 kPa
Sensitivity	8	1.9-4	2.6
Drained Shear Strength			
Internal Friction Angle (total stress)	3	7-13 deg	10.5 deg
Internal Friction Angle (effective stress)	6	26-37 deg	32.5 deg
Cohesion (total stress)	3	37-63 kPa	49 kPa
Cohesion (effective stress)	6	0-23 kPa	10 kPa
Pore Pressure Parameter (A)	9	+0.06 to +0.84	0.35
Brittleness Index	9	0-29%	9%
Consolidation			
Void Ratio	29	0.55-1.250	0.9
Compression Index (Cc)	14	0.10-0.60	0.23
Recompression Index (Cr)	18	0.002-0.06	0.02
Over-Consolidation Pressure	8	41-238 kPa	130 kPa
Over-Compression Ratio	8	1-4	2
Soluble Salts	13	1.8-9.4%	5.4%
Mineralogy: Very fine grained calcite comprises >90% of all samples. Additional minor phases include gypsum, detrital quartz and limestone fragments, salt and traces of clay minerals. Additional carbonate phases may also be present in minor quantities.			

Table G – Geotechnical properties of the Sedom Clay (after Adler and Harris, 1983)

consequence of prolonged ground motion. In view of its saturated condition, high void ratio, and the inevitability of interspersed sand laminae (especially in the proximity of wadi discharges), liquefaction of the Sedom Clay must be judged a distinct possibility when subject to immoderate earthquakes.

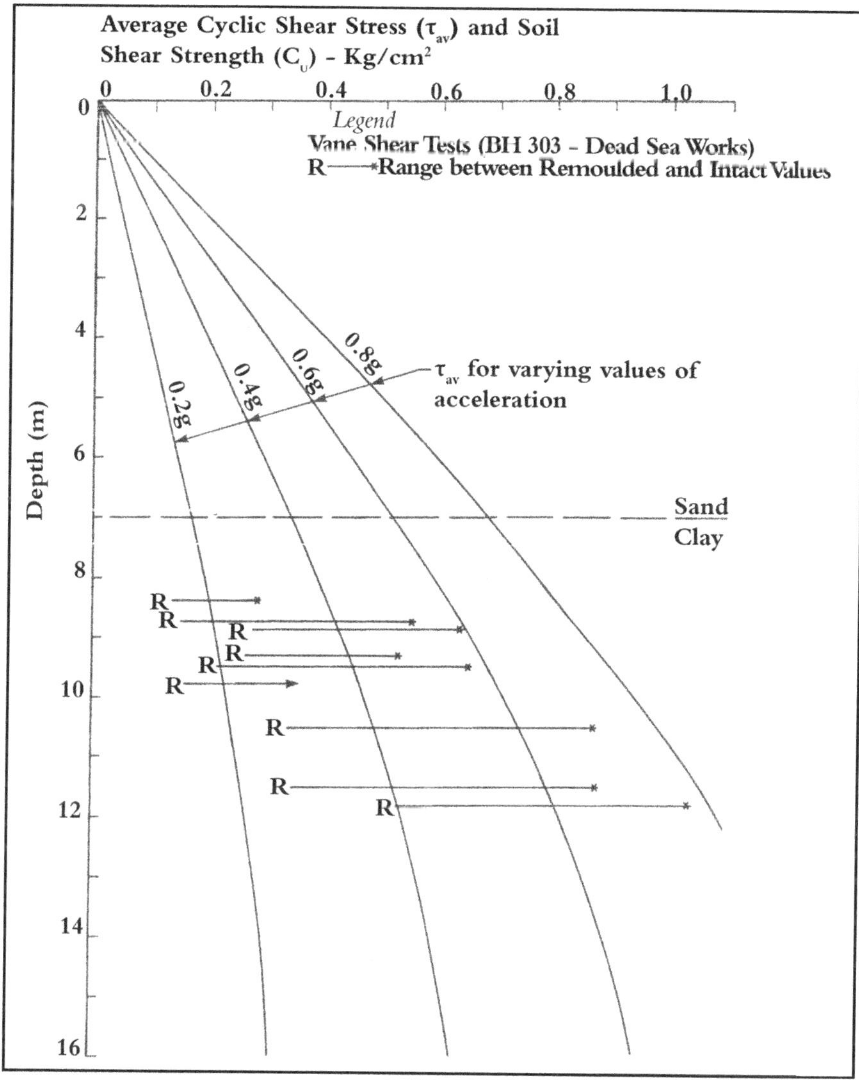

46. Induced cyclic shear stress for varying values of seismic acceleration for the Sedom Clay.

Youd and Hoose (1977) and Ambraseys (1988) both emphasise that shallow and saturated fluvial, deltaic and aeolian deposits of recent origin possess the greatest potential for liquefaction and ground failure. Such conditions would assuredly exist in the vicinity of wadi discharges and their associated floodplains. In this regard the postulated location of the legendary Vale of Siddim would constitute a potentially catastrophic geological setting. With a highly complex alluvial stratigraphy, underlain by a stratum of relatively weak micrite, major ground movements affecting both materials are a relative certainty when subject to seismic disturbance dependent, of course, on the magnitude and duration of the shock loading. The failure itself could be envisaged as being of a compound nature, the liquefaction of the Sedom Clay following the initial triggering of movement in the overlying alluvium.

Essentials for Liquefaction in the Vale of Siddim

There is no further reference in Genesis to the Vale of Siddim after the destruction of Sodom, Gomorrah, and the 'cities of the plain'. This suggests the Vale suffered the same cataclysmic fate, i.e. it was wiped off the map suddenly, which implies a massive loss of ground of such enormity that it swept everything into the Dead Sea to a watery grave – no survivors and no obvious remains!

For this to have occurred as a consequence of soil liquefaction, Drs Haigh and Madabhushi, of the Schofield Centre for Centrifuge Studies, Cambridge University, were retained by the BBC, London, to execute model centrifuge tests under the overall guidance of Dr Lynn Frostick, of Hull University, who specified the soil parameters of the in-situ soils required for modelling. Haigh and Madabhushi (2002) identified three essential requirements for liquefaction to be considered possible:

(1) The Dead Sea region must be seismically active.
(2) The soils adjacent to the shoreline must be liquefiable.
(3) Lateral spreading of the liquefied soil must be possible by means of an appropriate mechanism.

Cambridge University Centrifuge Tests

The Schofield Centre, Cambridge University, England, possesses one of the largest centrifuges in the world, and undertakes many challenging investigations into the ability of structures to withstand large earthquake forces. It is capable of accelerating a mass of one tonne through a radius of 4.125 metres with an acceleration of 150g. Such testing has become an essential preliminary to the construction of major structures, such as nuclear power stations and offshore drilling platforms.

The soils close to the edge of the Dead Sea reflect a seasonal variation of deposition. In wetter periods of the year sandier soils are deposited, whereas in drier periods the deposited soils are more of a silt-clay nature. The soils thus exhibit a degree of stratification, the sand strata being considered more liquefiable than the silt-clay strata due to their higher voids content, the silt-clay being more cohesive and less permeable. The model soils in the tests utilised various gradations of silica sand to simulate the actual soils. Figure 47 shows the model buildings in the tests, which were carried out on a ground slope of 1:20, or 3°.

The tests were successful, in that the sand became fully liquefied after being subjected to an acceleration simulating a representative earthquake (M=6), several earthquakes of this order having been registered in the Dead Sea region during

47. Model buildings on a slope in the Cambridge centrifuge tests (Haigh and Madabhushi, 2002).

the last century. From a study of seismic history this magnitude was considered a lower bound value with an acceptable return period. The conclusions reported by Haigh and Madabhushi (2002) were as follows:

> It has been shown that layered deposits of liquefiable sand and relatively impermeable silt, as found around the Dead Sea, are susceptible to liquefaction. It has also been shown that the presence of the impermeable layers slow the dissipation of excess pore-pressures and may hence greatly increase the magnitude of lateral spreading that would result owing to the formation of extremely low shear strength water films at the boundaries between layers.
>
> It is thus shown that liquefaction-induced lateral spreading after an earthquake of these layered deposits is a plausible explanation for the disappearance of the cities of Sodom and Gomorrah, with displacements of hundreds of metres or even kilometres being possible even on fairly gentle slopes.
>
> It is not possible to state categorically whether this is what happened, as insufficient evidence is available at this time, approximately 4000 years after the event is supposed to have taken place. In order to show this, archaeological evidence would need to be collected by searching for remains beneath the Dead Sea.

The purpose of the Cambridge centrifuge tests was to investigate the liquefaction potential of stratified Dead Sea sediments, composed of fine sand with embedded layers of silt. These constitute the bulk of shallow shoreline soils, and would have formed the greater part of the floodplain of the Vale of Siddim in which Sodom was located. Beneath the floodplain deposits is the Sedom Clay and, as has been mentioned earlier, it is sensitive enough to liquefy under favourable circumstances. Having been proved to extend to a depth of over forty metres at some locations, there is no reason to believe it would not have underlain the Vale of Siddim at depth, and was a component of the slide. As has been shown by analysis it would have been prone to liquefaction when subject to ground shaking by an earthquake possessing a minimum magnitude of M=6. The actual magnitude of the event is likely to have been considerably higher.

8. The Destruction

Archaeological Evidence

Any violent event leading to regional destruction on the scale implied by the dramatic events of Genesis, would have resulted in pandemonium among the inhabitants of those impacted. Wood (1999) writes:

> Evidence found at Numeira suggests the residents fled the town in haste. Most identifiable doorways from the latest occupation had been deliberately blocked. This apparently was an attempt to strengthen the homes against damage. In addition, no valuable small finds were discovered nor were there foodstuffs in the storage facilities. On the other hand, large quantities of pottery were found on the floors of houses, evidently too heavy and bulky to transport in the hasty evacuation. It appears the residents had some early warning, such as preliminary tremors, and did what they could to prepare. They shored up their houses, gathered up their valuables and as much food as they could carry, and fled their homes never to return. (Coogan, 1984: 80-81)

Donahue (1980, 1981) considered that there may have been two earthquakes which struck Numeira, the first occurring about 2400 BC, the second fifty years later, about 2350 BC, after which the settlement was abandoned. It is the latter earthquake which likely represents the seismic event responsible for the loss of Sodom. Donahue (1985) writes this about Numeira:

> It is suggested here that the tower collapse and extensive burn layers over the site were caused by an earthquake generated by fault movement.

Bab edh-Dhra would have been equally affected by any earthquake leading to the loss of Sodom, though unlike Numeira which was abandoned the settlement was reinhabited. It may be concluded, therefore, that Numeira lost its importance as a defensive outpost as a consequence of the earthquake, the industrial base of the Canaanite economy in the production of bitumen having been wiped out.

Excavations at Bab edh-Dhra indicate that during Early Bronze III the dead were interred in charnel houses constructed above ground, some of which covered a considerable area. The largest excavated was 15.5 x 7.8 metres (51 x 26 feet). The walls were constructed of sun-dried mud brick, the floor of pebbles, and the roof of reed matting daubed with mud and supported by wooden beams. The building had been severely damaged by fire, and among the remains were found charred beams and posts, bricks that had been scorched by excessive heat, funereal pottery and other allied objects, and many piles of scattered human skeletal remains amidst the ash and general chaos. Rast and Schaub (1980) described their excavations in the following terms:

> The extensive burn is clear evidence of the tomb's destruction by fire. Burning was concentrated along the interior wall in the center of both sections, where the majority of posts and beams were uncovered. Along the south wall impressions of desiccated beams angled down toward the interior transverse wall, indicating that they had collapsed in the center across the interior wall.

Wood (1999) quotes the following from those archaeologists who have intensively examined the ruins of Numeira where, surprisingly, remains of grapes were found interred among the ashes of the ruins:

> It is remarkable . . . that the grapes . . . were preserved even with their outer skins, due perhaps to the burning material which collapsed over the area and sealed these items. (Rast, 1981)
>
> Although carbonized whole grapes have been reported from Salamis, Hesban and Jericho, the size of the Numeira hoard, which consisted of over 700 whole grapes, is very uncommon. (McCreery, 1981)
>
> The infrequent small finds included more whole carbonized grapes with the stems attached and what preliminary analysis indicates were carbonized watermelon seeds (both evidence for dating the destruction of the site to late spring). (Coogan, 1984)

Neev and Emery (1995) have suggested a relationship between settlement in the South Basin of the Dead Sea and the fluctuation of water level. They concluded that the South Basin was settled about 3000 BC during Early Bronze I, and settlement continued without interruption until about 2200 BC towards the end of Early Bronze III, when drought conditions became exceptionally severe. Previously, the availability of an adequate supply of water had enabled them to survive periodic droughts, frequent earthquakes, and persistent hostilities.

With regard to the settlement of archaeological sites of Bab edh-Dhra and Numeira, the opinions of Neev and Emery differ from some archaeologists. They write:

> Bab edh-Dhr'a [Bab edh-Dhra] was occupied continuously between Early Bronze I and the end of early Bronze IV, whereas Numeira was inhabited

> only during Early Bronze III. . . . By the end of this age, both Bab edh-Dhr'a and Numeira (*Nimrim*, Isaiah 15:6) had been severely damaged by two earthquakes. Numeira was then deserted but Bab edh-Dhr'a was resettled immediately, perhaps by seminomadic people of the Intermediate Bronze age, who took advantage of the confusion caused by the natural disaster.
>
> Reconnaissance surveys by archaeologists who tried to find Early Bronze sites within the inner limits of the south basin and a short-term underwater archaeological survey in the Dead Sea (Baney, 1962) had negative results. Following these and positive results of their own excavations on land at the east fringe of the basin, Rast and Schaub do not expect additional Early Bronze settlements to be found within the subsiding part of the basin, including its south and west flanks. The possibility that remains of such sites could have disappeared because of geological processes – subsidence following earthquakes, submergence by the sea, and burial by sediments – was dismissed by Rast (1987) as speculative theories devoid of archaeological or geological support. Rast also concluded (after Klein, 1961) that tectonic activity was minimal during historic times.

Though the opinions of Neev and Emery, who are scientists, are at variance with those of Rast and Schaub, who are both eminent archaeologists, there would appear to be some merit in reconsidering this gulf of opinion. Early Bronze sites are distributed throughout the region, especially within areas possessing cultivable soils (Zori, 1962; Ben-Arieh, 1965). Also, there would have been settlements of a military nature strategically located along trade routes, or as observation posts on high points of ground, such as Numeira. The conversion of the South Basin into a sequence of evaporation ponds for the chemical industry curtails future archaeological exploration. However, as Figure 4 (Chapter 1) indicates, there have been many cultivable areas scattered around the shores of the South Basin in the past. Some of these likely supported smaller satellite settlements much further from the present shoreline. The fact they are now buried, unlikely to ever be discovered, does not prove they never existed. Neev and Emery (1995) sum up their views in the following terms:

> The results of archaeological exploration by both Albright *et al.* (1944) and Rast (1987) led them to suggest an association of the Sodom and Gomorrah tradition with Early Bronze III settlements in the south basin. Other sites of the same age and younger could have existed within that basin. . . . This is indicated both by biblical reports and by archaeological excavations along the west flanks of the basin, such as Mezed Gozal at the northwest corner of Mount Sedom where an eleventh century BC fortress was discovered (Aharoni, 1964a) and at En Gedi where a seventh century BC settlement and a Chalcolithic site were found (Mazar *et al.*., 1966). Other structures of the eighth century BC are along the northwest coast of the sea. (Bar-Adon, 1977)

The archaeological sites of both Bab edh-Dhra and Numeira offer evidence that both suffered considerable damage, likely caused by earthquake. The resultant chaos and destruction would have been catastrophic as, from all indications, it was also accompanied by fire. Professor Michael Finnegan, a forensic anthropologist with Kansas State University (BBC, 2001) examined three skeletons exhumed from Numeira. From the nature of the bone fractures, he concluded that the most likely cause of death was from being crushed by structural collapse of the building in which they were sheltering. Radiocarbon dating of the bones identified the year of death as about 2350 BC. The destruction of Numeira can, therefore, be judged contemporaneous to that of Bab edh-Dhra. Moreover, the date of 2350 BC is significant as this is when, according to Tubb (1998), there was a massive exodus of people out of Canaan, the cause of which is here concluded to have been caused by the destruction of Sodom and resultant devastation of the regional economy.

In the aftermath of the destruction, and with the mainstay of their economy ruined, any survivors remaining in the Dead Sea valley would have led a meagre existence compared to that prior to the event. Bewildered by the magnitude of the disaster, and unable to attribute its cause to anything other than godly wrath, the disruption of Sodomite civilisation would have been virtually complete. This would have had serious economic and political ramifications, not only within the Dead Sea valley itself, but as far afield as Egypt.

It would seem premature, therefore, to identify Sodom, Gomorrah, or any of the other notorious 'cities of the plain' with either Bab edh-Dhra or Numeira, especially as Wood (1999) states that excavations at Numeira indicate the site to have been in existence for less than about a hundred years. This would date the flourishing of Numeira to between about 2450-2350 BC, a period significantly less than that of the bitumen trade with Egypt, which can be dated back at least a thousand years earlier. As Numeira is situated upon a high point with good vistas to the southern approaches to the Dead Sea and westwards to the Judaean Hills, it is more likely to have served as a military observation post. Such a conclusion is supported by the fact that the area available to support a community at Numeira is somewhat restricted, totalling approximately two acres, despite the existence of thousands of graves in the immediate vicinity. With bitumen being the prime resource of the Sodomites, the location of Sodom would have been close to where the bitumen was harvested and processed. This means being located close to the water's edge. Therefore, the Sodomites would have depended upon warnings from outposts scattered throughout the region to give notice of any impending hostilities from advancing armies. Numeira, it is contended, would have represented one such outpost.

The foregoing argument cannot be extended to Bab edh-Dhra, a larger site encompassing about ten acres. Bab edh-Drha was discovered by William Albright and Melvin Kyle in 1924, and is sited closer to the postulated location of the Vale of Siddim. Albright (1933) described the location as follows:

> It is most emphatically not a city, but rather a temporary encampment, like Gilgal, or perhaps rather like the shadowy Baal-peor of Moab. Since the plain of Bab-ed-Dra [Bab edh-Dhra] is not suitable for cultivation, and is high above the gorge of the Seil ed-Dra, it would in any case be a very unsatisfactory place for a town. As a festival site for the inhabitants of the oases below it is admirable, since it is situated on the first convenient rise of terrain above the central oasis of el-Mezra'ah, some five hundred feet above the level of the Dead Sea.

Being located about eleven kilometres or so from the industrial metropolis of Sodom within the Vale of Siddim, a comfortable half-day's walk, Bab edh-Drha is most likely to have represented the festival site described in the Book of Jasher, where the inhabitants of Sodom and Gomorrah regularly congregated for their periodic revels. As the Book of Jasher is emphatic that 'all the people of Sodom and Gomorrah went there four times in the year' it is tempting to conclude that Sodom and Gomorrah were close neighbours. Also, one is inclined to think both communities may have lain within the Vale of Siddim, one a grim hive of industrial activity, the other, in contrast, a haven of residential tranquillity for the upper classes.

The 1960s witnessed a great deal of activity in the excavation of archaeological sites throughout the Near East. Between 1965 and 1967 Paul Lapp excavated Bab edh-Dhra. He showed it to possess features typical of the Early Bronze Age, and of a settlement that had been in existence for a long period of time. Subsequent excavations by members of the Expedition to the Dead Sea Plain (EDSP), a regional project funded by a number of agencies and private individuals dedicated to exploring that part of Jordan, showed Bab edh-Dhra to have been continuously occupied for about 1300 years. With such a long period of occupation it may be concluded that Bab edh-Dhra was also a permanent settlement as well as being a ceremonial site as previously thought by Albright.

Other sites in the same corner of the Dead Sea were identified as belonging to the same period. Rast and Schaub have been responsible for much of the pioneering work at these locations (Rast and Schaub, 1974; Rast, 1993; Schaub, 1991; Schaub, 1992). In total, five sites discovered, or visited, by Rast and Schaub have been shown to belong to the Early Bronze Age (3150-2200 BC). They are:

Bab edh-Dhra – from EBI to late EBIII and beginning of EBIV.
Numeira – EBIII.
Es Safi – EBI to EBIII.
Feifa – EBI to EBIII.
Khanazir – EBIII to EBIV.[1]

It is tempting to draw the conclusion that the five settlements found by Rast and Schaub are the five 'cities of the plain' as described in Genesis. However, in view of Strabo's claim 'that there were once thirteen inhabited cities in that region of which Sodom was the metropolis', it is impossible to link any of the

1. See Figure 4, Chapter 1.

above with those mentioned in the Bible. However, what is of interest is that all appear to have come to an abrupt end about the same time, with three of them having been consumed by fire. It is also relevant to consider Strabo's statement:

> [P]eople believe the oft-repeated assertions of the local inhabitants, that there were once thirteen inhabited cities in that region of which Sodom was the metropolis, but that a circuit of about sixty stadia [*11 kilometres*] of that city escaped unharmed; and that by reason of earthquakes and eruptions of fire and of hot waters containing asphalt and sulphur, the lake burst its bounds, and rocks were enveloped with fire; and as for the cities, some were swallowed up and others abandoned by such as were able to escape.

Despite the implication that the destruction was limited to a circuit of eleven kilometres or so according to Strabo, in actual fact it is likely to have been far more widespread. Bab edh-Drha is about eleven kilometres from the postulated Vale of Siddim, but at a high enough elevation to escape the effects of any ensuing tsunami following the earthquake. This may have influenced Strabo's statement. However, it is certain a tsunami following any such earthquake would have swept away all low-lying settlements in the Dead Sea valley far to north and south.

Regarding those settlements destroyed by fire, Hattem (1981) reports Rast and Schaub (1974) in the following terms:

> Of Bab edh-Dra – rubble remains of buildings are seen on the surface, and many door sockets and basalt mortars are strewn about. Of note, too, is the evidence of severe burning on many of the stones.
>
> Of Numeira – The site as a whole is covered with ashy soil, suggesting that it was not utilized after the destruction which appears so evident on the surface. On the north side especially, the soil is spongy ash, and can be picked up in handfuls.
>
> Of Feifeh – A surface feature worthy of attention was the thick spongy charcoal across much of the site. Feifeh is much like Numeira in this regard, the destruction dating probably to the end of the Early Bronze settlement.

Genesis 14 tells of Chedorlaomer's invasion of Sodom in the Vale of Siddim, how the vale was 'full of slimepits', and how the Sodomites who survived the battle 'fled to the mountain'. Of necessity the Vale of Siddim, as a commercial and industrial centre with a relatively large population, must have been well-watered. Wadi Kerak, Wadi Shuqei, and Wadi Hamed, collectively possess a catchment area of 400 square kilometres (see Table B, Chapter 2), and these flow into the embayment we observe today between the Lisan Peninsula and the eastern shore, (Bay of Mazra'a). This area must constitute a prime candidate for representing the legendary Vale of Siddim.

Thus, the Vale of Siddim would have constituted a deltaic plain, steadily expanding outwards into the waters of the North Basin by alluvium brought down in flood. This would have created a well-watered and potentially fertile area over the millennia of geologic time prior to human settlement. In view of its area, estimated as approximately forty square kilometres, several smaller satellites of Sodom may well have been sited in this vale, each clustered around the discharge points of the watercourses at the foot of the Hills of Moab. Whether any of the other 'cities of the plain' mentioned in Genesis, other than Sodom, were themselves located within the Vale of Siddim must remain unclear. The only city which is less difficult to identify is that of Zoar as it escaped unharmed, and was the refuge of the fleeing Lot that awful night. Therefore, Bab edh-Drha and Zoar must be concluded to be one and the same.

The Sodom Earthquake

Available bathymetric maps of the North Basin of the Dead Sea reflect a gentle erosion of the submarine fringes around much of the shoreline. This erosion pattern reflects the differing degrees of hardness in the alternating rock strata constituting the valley sides. However, immediately offshore to the postulated location of the Vale of Siddim there are a number of features suggestive of a massive submarine landslide. This may be premised from the presence of an ancient scar slope with an intriguing debris mound at its base, see Figures 48 and 49.

A consideration of this scar slope, and the offshore bathymetry, suggests the embayment forming the Bay of Mazra'a, between the Lisan Peninsula and the Jordan shore, was the result of a massive loss of ground, probably triggered by earthquake, as the Eastern Border Fault runs directly beneath the bay. The volume of displaced soil is conservatively estimated as being of the order of 3.5 cubic kilometres, i.e. 3.5 billion cubic metres, assuming the sea level was about −430 metres MSL at the time. The loss of shoreline would have been about six kilometres long.

Ben-Menahem *et al.* (1982) conducted a thorough statistical study of the earthquake history of Israel covering the last four thousand years, utilising the knowledge gained during the last century from advanced seismic instrumentation techniques. From this work a set of seismic risk maps was compiled with earthquake return periods of specific magnitudes. All the major fault systems were considered. With regard to the Dead Sea-Jordan Rift they indicated that for earthquakes having magnitudes equal to, or greater than M=6, the return period would be about 115 years, and for earthquakes having magnitudes equal to, or greater than M=7, the return period would be about 830 years. The last time a seismic event struck the Dead Sea with a magnitude M=6.0 was AD 1927, and for one with magnitude M=7.0 it was AD 1546. The work also concluded that the maximum possible magnitude of an earthquake within Israel was unlikely to exceed M=7.5.

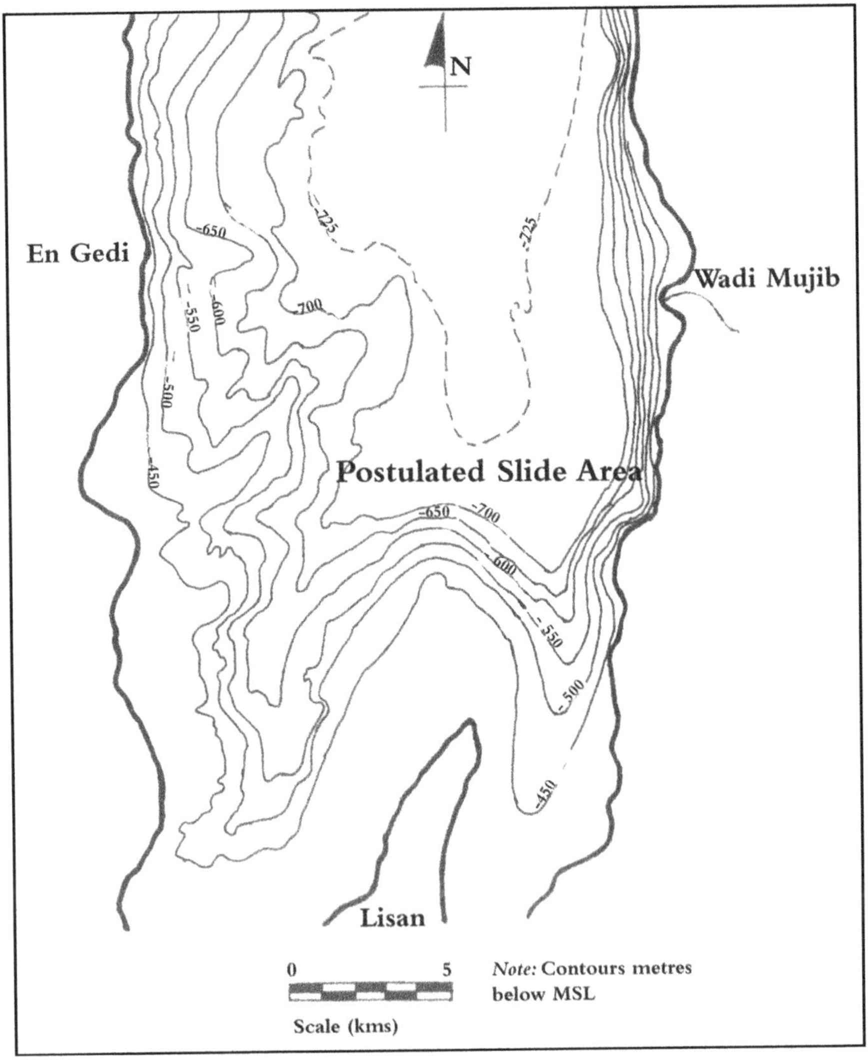

48. Bathymetric map of the southern part of the Dead Sea
(after Geological Survey of Israel, 2000).

Similar conclusions were reached by Husein *et al.* (1995) in their seismic mapping of Jordan, where most seismic events are related to tectonic movements within the Dead Sea fault system. Their research indicated that the highest ground accelerations and intensities were concentrated between latitudes 32.0° and 32.7°N on longitude 35.7°E. This is in the Jordan Rift midway between the Sea of Galilee and the Dead Sea, and over a hundred kilometres north of the Lisan Peninsula. However, as the Eastern Border Fault is a strike-slip fault its entire length is subject to high intensity earthquakes produced by the same grinding movements.

The forementioned work by Ben-Menahem *et al.* suggests the Sodom earthquake would have been associated with the Eastern Border Fault, and

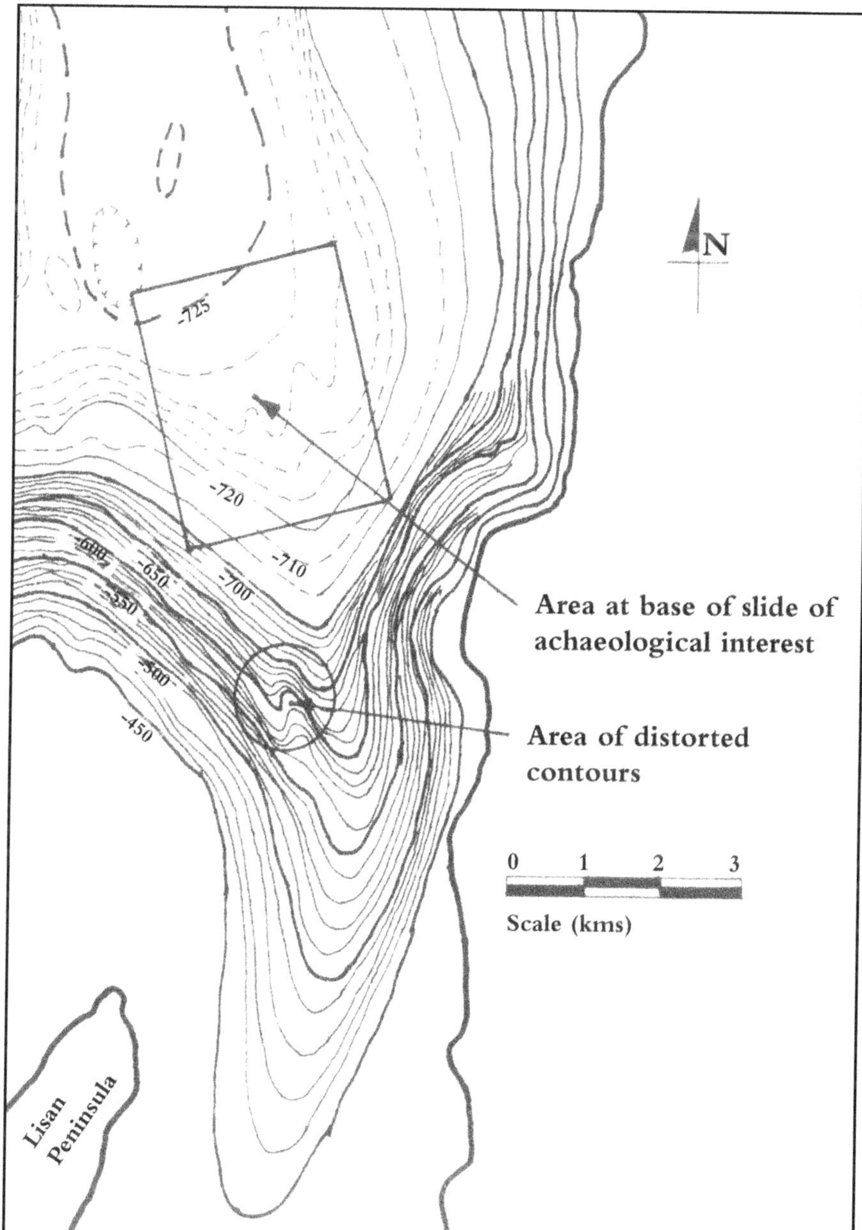

49. Scar surface beneath the Vale of Siddim indicating possible locations of archaeological interest.

likely to have possessed a magnitude greater than M=6.0, and one with a magnitude of M=7.0 would not have been unreasonable. Such magnitudes undoubtedly would have caused liquefaction in offshore sediment regimes in close proximity.

Ambraseys (1978), in his analysis of Middle East earthquakes, gives various relationships in which magnitude, intensity, epicentral radius, and ground acceleration, are interrelated. His relationship linking ground acceleration (a) in cm/s^2, and intensity (I) is given as:

$$\log a = 0.1 + 0.3\,I$$

Though Ambraseys considered this a weak relationship it permits a crude estimate of ground acceleration of about $0.7g$ for an earthquake intensity of at least IX on the Modified Mercalli Scale, an intensity surely warranted with the total loss of a city. Such acceleration, if prolonged, would certainly result in the liquefaction of essentially cohesionless saturated wadi floodplain deposits, and lead to the progressive failure of the underlying Sedom Clay.

In his inaugural Mallet-Milne Lecture at the Institution of Civil Engineers, London, Ambraseys (1988) emphasised that earthquake-induced liquefaction failures were confined to certain specific geological settings. They were invariably shallow events not extending deep beneath the ground surface, and were restricted to saturated, fluvial, deltaic and aeolian deposits. All these prerequisites would have been satisfied by the alluvial deposits of the Wadi Kerak underlying the Vale of Siddim, at least as far as initiation of the liquefaction process was concerned. However, once the top strata had become liquefied, and begun to spread downslope, the underlying strata would have become subject to the same liquefaction process, and they too would have started to fail and spread laterally. The entire event, therefore, can be envisaged as layers of liquefied soil progressively peeling off in rapid succession, until a stage was finally reached when little potentially liquefiable soil remained, or the shock tremors attenuated. This stage would have been reached, either when the intensity of shaking diminished to levels which could be resisted by the soil strengths, or the basin of sediments beneath the Vale of Siddim had been exhausted of liquefiable material.

Ambraseys reproduced an equation arising from work by Kuribayashi and Tatsuoka (1975) which, though based on Japanese data, suggested a means of making an approximate determination of the epicentral radius beyond which liquefaction would not normally be expected. With an epicentral distance (R_e), in kilometres, and earthquake magnitude (M), where M is on the Richter Scale, the relationship is:

$$\log (R_e) = 0.77\,(M) - 3.60$$

Though this equation has been based on Japanese data, there is no reason to believe a similar relationship would not apply to earthquakes in the Dead Sea valley. Therefore, it would seem reasonable to suggest that earthquakes with magnitudes ranging between M=6.0, and M=7.0, would result in an area of total liquefaction, measured from the epicentre, ranging from ten to sixty kilometres.

As the Vale of Siddim straddles the Eastern Border Fault the floodplain sediments of the Wadi Kerak would be exceedingly vulnerable to liquefaction.

Since liquefaction events are characterised by the sudden appearance of sand boils and mud volcanoes at surface, accompanied by differential ground settlement, followed by lateral spreading[2] the liquefaction of the Vale of Siddim would have been a terrifying ordeal for its inhabitants.

Because the various strata comprising a floodplain deposit possess differing strengths, not all strata would liquefy simultaneously. For liquefaction to be initiated, a minimum value of acceleration is required to overcome the inherent soil strength, and this is dependent on the intensity and duration of the seismic shock. But once the stratified deposit begins to break up decoupling between adjacent strata takes place, and huge slabs of more cohesive, i.e. clayey material, will be transported downslope amidst a sea of less cohesive sediment consisting of liquefied sand and silt. Only a modest slope is required for mass movement to occur.

The loss of the Vale of Siddim in such a manner would have been progressive. The first section to liquefy, and peel off beneath the waters, would have been the fringes of the plain directly bordering the sea. Then the landslide would have begun to eat backwards and downwards progressively as individual strata liquefied and began to move outwards. Amidst all this turbulence, structures and people would have been entombed in a morass of mud, and blindly swept away to a watery grave. Few would have escaped who were tardy in taking to higher ground.

It must be assumed that the epicentre of the Sodom earthquake was in close proximity to the Lisan Peninsula, the epicentral distance to the archaeological site of Numeira thus being about thirty kilometres. As mentioned earlier Donahue (1980, 1981) postulated that there may have been two earthquakes that struck Numeira, the first about 2400 BC, the second about 2350 BC, the latter being the greater of the two.

Attenuation of ground motion with increasing distance from an epicentre has been the focus of much seismic research (Seed and Idriss, 1982). For relatively high magnitude earthquakes, such as the Sodom earthquake, the peak horizontal acceleration likely to have been experienced at Numeira, would have been about 0.2-0.3*g*. This range equates to between VII and VIII on the Modified Mercalli Scale, i.e. considerable damage can be expected in otherwise well-built structures, but greater damage will occur to poorly built structures with ensuing collapse of walls. However, the damage would not be total. This is what can be observed at Numeira.

Though admittedly circumstantial, the available evidence does point to certain conclusions. Firstly, that a high magnitude seismic event associated with the Eastern Border Fault was responsible for the loss of Sodom and the liquefaction of the Vale of Siddim. Secondly, although it is overly ambitious to assign seismic values to such an earthquake, a consideration of all factors suggests a minimum intensity of at least IX, and possibly as high as the maximum of XII,

---

2. See Figures 42-44, Chapter 7.

which applies when destruction is total. Because the evidence lies beneath the waters of the Dead Sea a more positive assessment cannot be made. Overall, this suggests the Sodom earthquake to have had a minimum magnitude of between M=6.0 and M=7.0, with an approximate ground acceleration of at least 0.7*g* within the Vale of Siddim, leading to the loss of some 3.5 billion cubic metres of ground. Such a volume of displaced soil, as it slid beneath the waters, would, most certainly, have created a monstrous tsunami.

Landslide Scar and Debris Mounds

The bathymetric contours associated with the landslide scar surface, indicate the failure surface of the slide to have cut through alternating soft/hard strata much like an ice-cream scoop would cut through different coloured layers in a Neapolitan ice-cream, see Figure 49. Secondly, tantalising discontinuities to the bathymetric contours may be observed as 'tongues' protruding downslope (a) on the −510/530m intervals, and (b) on the −723m interval. These suggest accumulations of landslide debris. Also, and more intriguing, is the existence of three 'debris mounds' at the very base of the scar. It is these 'debris mounds' which, it is suggested, ought to be targeted by archaeologists, for they are more likely to contain the remains of any human habitation. The overall area of interest is approximately twenty square kilometres, though the individual mounds within the area present smaller and more specific targets.

Regrettably, the 'debris mounds' are located in some of the deepest water of the North Basin, in the vicinity of Longitude 35° 31'E and Latitude 31° 26'N. Though this location lies within Jordanian territorial waters it is close to the International Border with Israel, and some overlap of the border by the debris plume would be expected. As a consequence, the accessibility for archaeological research poses a unique challenge.

What can be expected to be found? After more than four millennia beneath the waters of the Dead Sea, any remains of Sodom are likely to be shrouded with thick accumulations of salts of varying composition, but mostly of gypsum. However, a tsunami event resulting from any liquefaction induced landslide would have two major effects. Firstly, the tsunami runup would deposit debris on both shores leaving onshore signatures in the form of strand lines. These would constitute positive evidence of the event having occurred. Such strand lines are likely to be at variance with normal raised beaches and, as has been noted earlier, a reassessment of radiocarbon dating on wood fragments may also prove useful. Secondly, the failure would have covered lake bottom sediments with a thick layer of fresh material in which would be entombed any remains. As the volume of slide debris is substantial and dispersed widely over the lake bottom, the use of modern geophysical tools in any search would be a necessity.

Since most of the dwellings would have been of a humble nature, probably of dry stonewall construction, they would have crumbled in the earthquake. Remains of these, therefore, are unlikely to be found. Only the remains of temples, shrines, or palaces would likely still exist as these would be of more substantial construction, i.e. possibly carved stone or interlocking masonry.

As high salinity is often beneficial to the preservation of organic remains by inhibiting bacterial decay, it is possible human or vegetative remains might eventually be recovered. If so they would be invaluable, enabling the demise of Sodom to be dated more accurately than present considerations permit.

Geology of Delta Formations

The identification of an ancient submarine landslide associated with the deltaic sediments of the Wadi Kerak, prompts comparison with the offshore sediments of Wadi Mujib, also known as the Arnon River. The Wadi Mujib enters the North Basin of the Dead Sea approximately eighteen kilometres north of the postulated location of the Vale of Siddim, now the Bay of Mazra'a.

The catchment area of Wadi Mujib is considerably less than that of Wadi Kerak, being 6.7 square kilometres; over twenty times smaller. Its length is about seventy kilometres, and it follows a tortuous path out of the mountains. As with Wadi Kerak its flow is ephemeral, though the flow of the Kerak benefits from contributions from more reliable aquifers with sources in sandstone. A comparison of the two deltas is not unreasonable as the sediment loads carried are of similar origin, though disparate in volumes carried.

Harris and Partridge (1980) summarise the formation of deltas in the following terms:

> The formation of a delta represents the continuing ability of a river system to supply and deposit sediments more rapidly than they can be removed by offshore wave and current action. Numerous factors affect this sedimentation process, these include the lithology of the rocks in the river basin subject to erosion, the climate, rainfall, prevailing offshore currents, tidal range, etc. The most important of these is possibly climate as this controls both the rate at which water is supplied to the drainage area, and thus the quantity of sediment which is transported to the delta region. The latter may be divided into bed load, which consists of coarser particles which travel close to the bed of the channel, and suspended load, which consists of finer particles in suspension, the hydraulic conditions being unsuitable for deposition. Channels carrying large volumes of bed load are normally wide and shallow in comparison to those carrying a high percentage of suspended load.
>
> The accumulation of deltaic deposits takes place in a generally orderly fashion. The first materials to be deposited at an offshore location are the finer sized materials (known as the bottomset beds of proto-deltaic facies). These are generally clay and silt-sized materials, deposited at depths where transportation velocities are low enough to permit deposition. Above these, as the delta grows outwards in an offshore direction, increasingly coarse bar finger materials are deposited (known as foreset beds).... Above the foreset

beds there are once again a series of finer materials (known as topset beds), which are representative of finer flood-plain deposits.

Dependent on transportation velocity of the river, and the relative proportion of bed to suspended load, is the ability with which a channel can become blocked. Rivers which are subject to flash floods, and which remove and transport vast quantities of sediment in short periods of time, can be subject to rapid blocking of their channels when velocities are suddenly checked.

Wadi Kerak and Wadi Mujib are both subject to flash floods, as indeed are all watercourses entering the Dead Sea. The sediment load carried by such streams running in flood is often sufficient to overwhelm and block adjacent roads. The effects of such flash flooding may be appreciated (Figure 22, Chapter 2).

The basis for comparing the profiles of the two deltas (Figure 50 – depth and distance scales not shown) relies on the latest version of the only available bathymetric map of the Dead Sea (Geological Survey of Israel, 2000). Though there are shortcomings in the information depicted, it is the only one in the public domain and, therefore, there is no better alternative. Despite the undoubted defects in the depiction of the bathymetry certain conclusions may be reached in a comparison of the two delta profiles.

The delta profile of the Wadi Mujib indicates it to possess a classical 'S' shape when viewed in cross-section. Close inshore the surface gradient is about 1:5 (12°), flattening progressively and uniformly with increasing distance from the shoreline to assume a gradient of about 1:10 (6°) at elevation −660 metres. Below that elevation the gradient diminishes further to become less than1:50 (about 1°). There is nothing unusual about the profile of the delta of Wadi Mujib to suggest it may have been subject to seismic disturbance at any time in its history.

The same cannot be said about the delta of the Wadi Kerak. The profile of the delta is substantially different. At shallow water depths, i.e. above elevation −440 metres, the gradient is fairly uniform at about 1:50 (1°), then decreases abruptly to about 1:100 (0.5°) at elevation −460 metres, before increasing progressively with increasing depth. At an elevation of −660 metres the gradient is about 1:10 (6°), which is identical to that of the delta of Wadi Mujib at the same depth. Below this level the two profiles assume similar gradients.

The delta of Wadi Kerak possesses a pronounced convexity in its profile at higher elevations, and this is considered indicative of a reconstitution of the original floodplain as sedimentation expands into deeper water. Since over four millennia have elapsed since the Sodom earthquake it would be logical to expect the wadi to resume its natural geomorphological pattern of sediment deposition.

It is unlikely the loss of the Vale of Siddim as a consequence of seismic liquefaction (c. 2350 BC) was the only instance of such a failure in the course of the long geological history of the Dead Sea. With lengthy return periods of high intensity earthquakes, it is feasible the same location may have been subject to previous failures well before the region became inhabited.

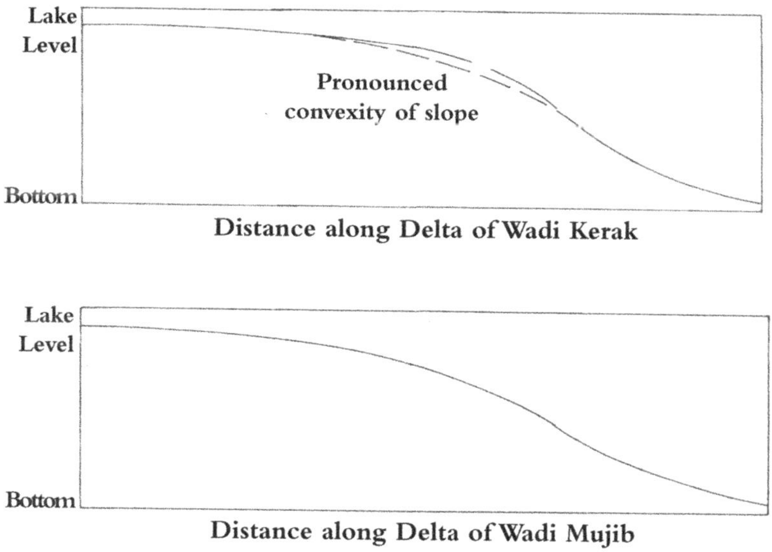

50. Comparative profiles for the deltas of Wadi Kerak and Wadi Mujib.

The Tsunami

Frequently, strong submarine earthquakes are followed by tsunamis when tectonic activity results in a displacement of the sea floor, or when landslides lead to displacement of large volumes of water. Consequently, where no relative displacement occurs, for instance in strike-slip movements, there is no tsunami. The celebrated Lisbon earthquake of 1755, one of the most destructive on record, originated in a massive subsidence of the sea floor. The sea at Lisbon fell immediately after the first shock wave was felt, but withdrew only to return as a gigantic wave about twelve metres high. The ebb and flow of the sea continued for some time, and was of such severity that the water level in lakes as far away as Scotland oscillated up to a metre high for an hour after.

One of the more recent tsunami events is the one which struck the east coast of Japan, March 2011. In this case the earthquake's epicentre was located offshore Sendai, and struck with a magnitude of 9.0. The damage due to the earthquake alone, though extensive if considered in isolation, was dwarfed by the wider chaos and havoc caused by the ensuing tsunami, the effects of which were felt on the far side of the Pacific Ocean, where $10 million of damage was inflicted on harbours in California and Oregon.

When tsunamis are produced by landslides, as opposed to tectonic activity resulting in shifts to the sea floor, the heights of the tsunami waves generated can be immense. Such events are termed *megatsunamis*. That which followed the collapse of Mount St. Helens, Washington, USA, in 1980, when part of the mountain detached and surged into nearby Spirit Lake, is reported to have attained a height of 260 metres. A parallel event, within living memory, but of lesser magnitude, was the landslide in the reservoir of the Vaiont Dam, Italy, that occurred in 1963. The reservoir was long and narrow and the loss of the left bank immediately upstream of the dam resulted in two tsunami waves being generated. The first wave overtopped the dam, the second travelled to the far end of the reservoir, then returned to overtop the dam a second time. Fortunately the dam remained intact, experiencing only minor damage, but the town of Vaiont downstream was virtually destroyed. Strangely, the church was barely touched.

A more impressive tsunami, but one which resulted in significantly less damage, was when 40 million cubic metres of rock and ice were dislodged and plummeted into Lituya Bay, Alaska, in 1958, following a magnitude 7.7 seismic event. The resultant wave was 524 metres high, the highest in recorded history. However, there is scientific evidence for some prehistoric events that dwarf these. The *megatsunami* arising from the asteroid that struck the Gulf of Mexico, off the Yucatan, some 65 million years ago, and which is considered to have terminated the Age of the Dinosaurs, is believed to have been as high as three kilometres.

The description of the fate of Sodom given by Strabo is highly supportive of the fact that the destruction was due, at least in part, to a tsunami wave following the earthquake. Strabo writes:

> by reason of earthquakes and of eruptions of fire and of hot waters containing asphalt and sulfur, *the lake burst its bounds*, and rocks were enveloped with fire; and as for the cities some were swallowed up and others abandoned by such as were able to escape [*the italics are mine*].

The translation of a lengthy piece of shoreline six kilometres or so into the waters of the Dead Sea following the seismic event envisaged would, undoubtedly, result in the creation of a tsunami of some immensity. The wave, so generated, would have travelled rapidly outwards from its source, probably breaking into two secondary waves once the primary wave struck the opposing western shore. The situation is somewhat analogous to sitting down quickly in a bath of water – the displaced water rockets up to the far end of the bath, reflects, returns, and eventually subsides.

The geography of the Dead Sea basin suggests that one wave, the larger of the two, would have travelled northwards encompassing the entire length of the North Basin before being reflected from its farther shore. In fact there could be several reflections from both ends of the Dead Sea Basin before the disturbed waters abated. The second wave, significantly smaller, would have passed through the constriction of the Lynch Strait before spreading out across the salt playa, or shallow waters of the South Basin, before being reflected in its turn.

Estimating the height such a tsunami wave might have attained is fraught with uncertainties. Depending on the assumptions made, such a wave may have attained a height of between 15 metres and 150 metres. Such estimates are too crude to be of much value, except to confirm that the loss of 3.5 billion cubic metres of soil as a result of the slide would have generated a tsunami wave of frightening proportions to the inhabitants, and of fearsome aspect.

However, such uncertainties present a unique opportunity to a university with a suitably equipped hydraulics laboratory, in constructing a scale model of the Dead Sea basin, reproducing the postulated landslide, and recreating the resultant tsunami. The conclusions of such a modelling study would be an invaluable contribution to the better understanding of the Sodom event. The elevation attained by the tsunami that lashed the Dead Sea would add credence to the understanding of why Biblical Zoar was spared, and more positively identify it with present-day Bab edh-Dhra.

The present elevation of Zoar sets an upper bound for the maximum elevation of any tsunami wave generated following the earthquake. As the tsunami ran up the shelving shorelines at the northern and southern extremities of the Dead Sea, so its height would have increased commensurately, before its energy became dissipated. It would have demolished everything in its path. How far it may have travelled up the Jordan and Arava valleys is open to speculation at the present, but the results of such a study, as mentioned above, would prove conclusive.

Albright (1933) showed the archaeological site of Bab edh-Drha to be 500 feet (150 metres) above the level of the Dead Sea at the time of his discovery. Dead Sea levels have since fallen, and presuming Bab edh-Drha to represent Biblical Zoar, its height at the time of the earthquake would have been approximately 200 metres above the Vale of Siddim.

Niemi *et al.* (1997) reported finding an extensive submarine slide in the delta of the Jordan River located 5.5 kilometres offshore. It was discovered using seismic reflection techniques and is estimated to have possessed a volume of about two million cubic metres, and took place on a slope of less than 1° at a depth of 250 metres. The slump area was covered by a thin veneer of recent sediment. From eye-witness accounts of 1927 the slide was concluded to have been triggered by the Jericho earthquake (estimated magnitude 6.2) which occurred at the same time. The eye-witnesses described a wave rising from the depths of the lake, spreading rapidly towards the northern shore, and piling up water about a metre high.

James McCalpin (2009) reports the same event declaring it to represent a tsunami, and stating that 'eight other slides in the same location suggested a repeat time of significant earthquakes on the Dead Sea Fault of several thousand years over the last 20-30 ka'.

A tsunami wave one metre high, arising from a submarine slide of two million cubic metres, would be puny compared to that created by a liquefaction slide involving a volume two thousand times greater, such as that estimated to have resulted from the Sodom earthquake. The former would have merely lapped the shore in comparison, whereas the latter would have devastated the low-lying valleys to north and south, penetrating far inland. The possibility of a past tsunami event of such magnitude, arising from such a cause, appears to have been neglected by past researchers. It is hoped this apparent oversight will spur scholarly endeavour in this direction.

Transformation of Lot's Wife

A surprise to many first-time visitors to the Dead Sea is to observe what appears to be icebergs floating upon its waters from time to time, particularly towards its southern extremities. This initial impression is corrected by common sense, for they are merely blocks of salt.

There is no reason to believe that such blocks of salt were not as prevalent in antiquity as they are today, especially when lake levels were significantly lower and salt content higher, as may be implied at the time Sodom existed. The translation of a lengthy piece of shoreline beneath the waters, resulting in a tsunami leading to a reflected tidal wave, could well have left blocks of salt stranded at very high elevations on the shorelines, considerably above lake level. If such a block of salt had been stranded anywhere close to the path followed by Lot and his daughters on their uphill flight to Zoar, human incredulity and superstition would have elevated the disappearance of his wife into current legend.

The Saunders Mini-Sub Search

In 1999, Michael Saunders, on behalf of the Mysteries of the Bible Research Foundation of Irvine, California, carried out some diving in the North Basin using a mini-sub in an attempt to locate the remains of Sodom and Gomorrah. His campaign was based upon anomalies that had been detected on satellite images of the Dead Sea, from which it was concluded these might represent the underwater remains of buried cities. All the locations identified were towards the northern extremity of the North Basin, close to the discharge of the Jordan River. Saunders only managed four dives, and these were restricted to Israeli waters due to military sensitivities.

Whether any of the locations were identified with one or more of the 'cities of the plain' is uncertain. But comparing the bottom elevations, where these anomalies were noted, to the historical variation of Dead Sea water levels as inferred by Neev and Emery (1967), it may turn out that the sites identified by Sanders could be Neolithic or Chalcolithic, a much earlier age to the Early Bronze which was contemporary with Sodom.

Jonathan Petre writes of the Saunders expedition in The Sunday Telegraph of 26 March, 2000 as follows:

> Mr Saunders is in little doubt that the salt-covered mounds, found over an area 800 yards square, are man-made structures. He said 'I have spoken

to geologists and nobody has come up with the suggestion that they are natural phenomena. We don't know what else they could be if they are not ruins. But we need more conclusive evidence by chipping off the salt. That's why we need to go back.'

Though the expedition did not fulfil its intended objective, it produced the first underwater images of the bottom of the Dead Sea, showing it to be a vast accumulation of gypsiferous and other saline growths. Saunders described the bottom to be remarkably flat in the area he explored.

With Saunders determined to return to the Dead Sea in the future it is hoped his underwater reconnaissance will be located further south to investigate the 'debris mounds' at the foot of the landslide scar off the Bay of Mazra'a. It is unfortunate this is located close to the international boundary with Israel. Any diving expedition is, therefore, likely to encounter similar obstacles to those which caused the curtailment of his 1999 programme.

Future Searches

Even if a future diving expedition was to receive the approval of both governments of Israel and Jordan, it is unlikely much useful information would be gleaned at the present time, apart from underwater images which by themselves might prove tantalising. An essential preliminary to a search standing any hope of success, would be to obtain an accurate map with contours of the sea bottom at close intervals in selected areas of interest. There are a host of geophysical tools readily available which might be considered for conducting such surveys. This would enable the identification of specific sites of interest prior to committing a full scale archaeologically directed assault.

One of the tools that might be considered for preliminary surveying is Continuous Refraction Seismic Profiling (or CRISP). This technique utilises a bottom drag sled equipped with hydrophones and an acoustic energy source activated through an umbilical attachment from the towing vessel, which receives and processes the signals received. It is claimed that the refraction technique possesses a number of advantages over the previously favoured reflection method of seismic profiling.

Another method is Side-Scanning Sonar, where a sonar 'fish' is towed above the sea bottom. Again, transducers transmit narrow pulses of sound at right angles to the towing vessel's track, and the reflected signals are recorded. The resultant sonar images can yield valuable information on the underwater surface topography and its composition, especially regarding the presence of sea bottom anomalies.

The geophysics involved could, with advantage, be combined with the exploration for hydrocarbon reserves. Yossi Lingojsky (2000), in relation to the success of such methods for determining offshore gas fields in Israeli waters of the Mediterranean, writes:

> Identifying these gas reserves isn't a stroke of incredible good luck. Extensive work to probe the riddles precedes the drilling. The detective work has become astoundingly precise, producing the 100 percent success ratio.
>
> While only actual drilling can confirm definitively that a gas or oil reserve exists at a particular site, the predictive accuracy of contemporary geophysical methods developed in recent years has been very impressive.

This attitude indicates geophysics to have considerable application in overcoming the challenge of determining the whereabouts of the remains of Sodom and its satellite cities. The field of remote sensing is an expanding one, with newer tools being invented as scientific knowledge expands. There is little doubt that if the specific technology to overcome the challenge does not exist currently, sooner or later it will become available.

The Aftermath

Doomed by virtue of its geography, the tragedy that overwhelmed Sodom may be concluded as a geological disaster waiting to happen. The floodplain of the Wadi Kerak, constituting the Vale of Siddim, extended further and further into deeper and deeper waters throughout the preceding millennia. Eventually, it reached a point when ground-shaking, as a result of seismic activity, caused the unstable sediments of which it was composed to collapse like a house of cards. As a result the liquefied mass began to flow and disappeared beneath the waters.

On that awful night, it probably took no more than about twenty minutes after the first tremor of the earthquake was felt, to when the angry waters of the ensuing tsunami began to abate. The only evidence that remained on the rising of the sun next day would have been the fringe of destruction around the shorelines. Sodom, and its satellite cities with all their inhabitants, had been swept from the face of the earth. The Dead Sea would have been transformed into a dead place with an evil reputation. Passed down through legend, the destruction of Sodom by such means may, perhaps, have contributed to its present name.

Undoubtedly, the news of the catastrophe would have rapidly travelled far and wide, and it may be expected in those days, when superstition was rife, that numerous causes would have been attributed to the event. Featuring among these would be that of an angry god or gods, and a legacy of myth left to mystify future ages. The magnitude of the disaster would have had a profound impact, sufficiently great for the story to be woven into the Old Testament as we know it today.

Mass emigration out of the region would have ensued, as the survivors fled to escape the dire consequences of any repetition of the disaster, and many of these would have sought refuge in Egypt. We know that during the First Intermediate Period in Egypt large numbers of Semites began settling in the Nile delta region, and their numbers mushroomed later during the Twelfth and Thirteenth Dynasties. However, a steady influx of Canaanites would have created economic and political uncertainty within the host nation, and it is likely many refugees would have been enslaved as a result. Perhaps these were to become known as Hebrews and, several centuries later, to be led out of Egypt by Moses in the Exodus.

There is a variation of opinion among scholars as to the date when the Old Kingdom of Egypt ended, and the First Intermediate Period began. It is generally concluded as between 2400 BC and 2200 BC. This brackets the conclusions expressed herein that the destruction of Sodom occurred c. 2350 BC.

Davidovits and Morris (1988) have indicated that after the Old Kingdom had come to a close the quality of pyramid construction declined. Though this decline may be attributed to the loss of techniques, it is equally possible to conclude the decline arose through the shortage, or lack, of high quality materials that had their origins in the Dead Sea valley. If this is so, then it lends substance to the suggestion that the Sodomites may have been among the first chemical engineers to practice their art.

9. Conclusions

For the most part, planet Earth is a habitable place for humankind to live and, we are led to believe, possesses a far more clement environment than any of our sister planets. However, on occasions extreme events occur, such as earthquakes, volcanic eruptions, fires, floods, storms and tempests, and our peaceful existence is disrupted for a period. Science has attempted to understand why such excesses occur, and how to limit the ensuing disasters when we cannot control or avoid them. The study of case histories serves a very useful purpose in this process of understanding the 'whys and wherefores'. The destruction of Sodom is a case in point as the myth and legend surrounding it has persisted through the millennia.

Roger Ascham (1515-68), a learned Elizabethan, wrote – 'By experience, we find a short way by a long wandering.' Having reached this point it is feared readers may be tempted to believe the author has, indeed, led them on an overlong and tortuous path. To the extent these wanderings may be deemed excessive is left to the reader. However, it is hoped these wanderings, and the inevitable digressions along the way, have not been accompanied by excessive tedium. The sage words of Melvin Kyle, recounted at the beginning of this book, regarding the retelling of dry-as-dust facts, has been uppermost in the author's mind throughout.

It is hoped that what has been presented, will be accepted as a contribution from a technical perspective to that enigmatic question – what happened to Sodom, Gomorrah, and the other 'cities of the plain' on that fatal occasion so long ago? Therefore, let us summarise the findings as they relate to the issue:

(1) Far from being located beside, or upon, the arid salt pans of the South Basin, the city of Sodom was a thriving centre of trade and industry situated in the fertile Vale of Siddim located northeast of the Lisan Peninsula and occupying the sedimentary floodplain of the northern distributary of the Wadi Kerak. The economy of Sodom was based largely on the exploitation of Dead Sea bitumen and its export to Egypt and beyond.

(2) The destruction of Sodom is concluded as having taken place about 2350 BC with devastating impact on the economies of Canaan and Egypt, which were closely interlinked. It is possible the raw resources of Canaan, particularly bitumen, were of greater importance to the economy of Egypt than hitherto realised.

(3) The dating of the event coincides with the end of Early Bronze III. This marks the beginning of the decline of the Old Kingdom in Egypt, and at the same time is coincident with mass emigration out of Canaan. To what extent the sudden termination of the bitumen trade between Canaan and Egypt is directly responsible for any civil strife, societal disruption, or cultural animosity between the host nation and this influx of immigrants must be left to the attention of scholars.

(4) Geotechnical reasoning concludes the Vale of Siddim to have been lost as a result of seismic shock loading on the Eastern Border Fault of the Dead Sea. The fault directly underlies the area. The loss of ground involved amounted to about 3.5 billion tons of soil, and constituted the alluvial plain of the Wadi Kerak. The weak alluvial and calcareous sediments beneath were subject to liquefaction, and slid beneath the waters.

(5) The submarine slide would have initiated a massive tsunami, and this would have inundated low-lying territory bordering both basins of the Dead Sea. Only settlements on higher ground would have escaped the tsunami. The archaeological site of Bab edh-Dhra, believed to represent the Biblical city of Zoar, was one spared. Lower lying settlements would have been wiped out by a combination of earthquake, fire and flood.

(6) Though the prosperity of Sodom and its satellite cities arose primarily from the harvesting of bitumen, its refinement, and its export to Egypt, it is likely other resources were also exploited. With the help of Egyptian expertise in the art and practice of chemistry, there is no reason to believe the Sodomites were not some of the world's first chemical engineers in manufacturing a host of products with a ready market in Egypt. Sodom would have been a hive of industry, as is the region today.

(7) The Biblical account of the destruction of Sodom and its satellite cities is largely substantiated by this scientific commentary. The writers of Genesis had little knowledge of seismology, less knowledge of geology, and were at a distinct disadvantage compared to ourselves in rationalising the catastrophe that overwhelmed Sodom and its neighbours on that fatal occasion in c. 2350 BC. The Biblical account is exactly what one would expect from untutored observers in the dawn of history, passed down through the ages by oral tradition.

Bibliography

Abraham H., *Asphalt and Allied Substances*, Van Nostrand, New York. (1929)

Adler, S. and Harris, G.M., 'The stockpiling of carnallite upon Dead Sea alluvium', *Proceedings 7th Asian Regional Conference on Soil Mechanics and Foundation Engineering*, Haifa, Israel. (1983)

Agricola G., *De Natura Fossilium*, Per H. Frobenium and N. Episcopium, Basle. (1546)

— *De Re Metallica*, Per H. Frobenium and N. Episcopium, Basle. (1556) Dover Publications, New York. (1950) Translation by Hoover and Hoover.

Aharoni, Y., 'Mezad Gozal', *Israel Exploration Journal* 14: 112-13. (1964)

Albright, W.F., *The Archaeology of Palestine and the Bible*, Revell, New York. (1933)

Albright, W.F., Kelso, J.L. and Thorley, J.P., 'Early Bronze Age Pottery from Bab-ed-Dra in Moab', *Bulletin of the American School of Oriental Research*, vol. 95, pp. 3-13. (1944)

Alonso-Azcárate, J., Bottrell, S.H. and Tritlla, J., 'Sulfur redox reactions and formation of native sulfur veins during low grade metamorphism of gypsum evaporites, Cameros Basin (N.E. Spain)', *Chemical Geology*, vol. 174, pp. 389-402, Elsevier, Amsterdam. (2001)

Alter, R. and Kermode, F., Introduction to *The Literary Guide to the Bible*, Harvard University Press, Cambridge, Massachusetts. (1987)

Ambraseys, N.N., 'The Middle East – A reappraisal of seismicity', *Quarterly Journal of Engineering Geology*, 11: 19-32. (1978)

— 'Engineering Seismology', *Earthquake Engineering and Structural Dynamics*, 17:1-105. (1988)

Arkin, Y. and Gilat. A., 'Dead Sea sinkholes: an ever-developing hazard', *Environmental Geology* 39(7):711-722. (2000)

Avi-Yonah, M., *The Madaba Mosaic Map*, Israel Exploration Society, Jerusalem, pp. 1-80. (1953)

Avni, R., Bowman, D., Shapira, A. and Nur, A., 'Erroneous interpretation of historical documents related to the epicenter of the 1927 earthquake in the Holy Land', *Journal of Seismology*, 6:469-476, Kluwer, Netherlands. (2002)

Baney, R.E., *Search for Sodom and Gomorrah*, Christian Approach Mission Press, Kansas City. (1962)

Bar-Adon, P., 'Another settlement of the Judean Desert Sect at Ein el-Ghuweir on the shores of the Dead Sea', *Biblical Association – Bulletin of the American Schools of Oriental Research* 227: 1-25. (1977)

Barsoum, M.W., Ganguly, A. and Hug, G., 'Microstructural Evidence of Reconstituted Limestone Blocks in the Great Pyramids of Egypt', *Journal of the American Ceramic Society*, vol. 12. (2006)

Bar-Yosef, O., 'Prehistory of the Jordan rift', *Israel Journal of Earth Science*, 36:107-19. (1987)

Begin, Z.B., Ehrlich, A. and Nathan, Y., *Lisan Lake, the Pleistocene Precursor of the Dead Sea*, Geological Survey of Israel (Bulletin 63), Jerusalem. (1974)

Begin, Z.B., Nathan, Y. and Ehrlich, A., 'Stratigraphy and facies distribution in the Lisan Formation – New evidence from the area south of the Dead Sea, Israel', *Israel Journal of Earth Sciences*, 29:182-189. (1980)

Begin, Z.B., Broecker, W., Buchbinder, B., Druckman, Y., Kaufman, A., Margaritz, M. and Neev, D., *Lake Lisan levels over the past 30,000 years. A preliminary report.* Report G.S.I. 29/85, Geological Survey of Israel. (1985)

Ben-Arieh, Y., *The Central Jordan Valley* (in Hebrew), Hakibbutz Hameuchad Publishing House, Tel Aviv. (1965)

Ben-Menahem, A., 'Earthquake catalogue for the Middle East (92 BC-AD 1980)', *Bollettino di Geofisica Teorica ed Applicata*, 21:245-310. (1979)

—'Variations of slip and creep along the Levant rift over the past 4500 years', *Tectonophysics*, 80: 183-87. (1981)

Ben-Menahem, A., Nur, A. and Vered, M., 'Tectonics, seismicity and structure of the Afro-Eurasian junction – the breaking of an incoherent plate', *Physics of the Earth and Planetary Interiors*, v12:1-50, Amsterdam. (1976)

Ben-Menahem, A., Aboodi, E., Vered, M. and Kovach, R.L., 'Rate of seismicity of the Dead Sea region over the past 4000 years', *Physics of the Earth and Planetary Interiors*, 14: 17-27. (1977)

Ben-Menahem, A., Vered, M. and Brooke, D., 'Earthquake risk in the Holy Land', *Bollettino di Geofisica ed Applicata*, 24: pp. 175-203. (1982)

Bentor, Y.K., 'Geological events in the Bible', *Terra Nova*, Blackwell Scientific 1(4):326-338. (1989)

Ben-Tor, A., 'The trade relations of Palestine in the Early Bronze Age', *Journal of the Economic and Social History of the Orient*, vol. 29, pp. 1-27. (1986)

— 'On the nature of Egyptian presence in Eretz-Israel in Early Bronze I', in *Jerusalem* (ed. E. Israel); Yadin Volume, pp. 31-36, Israel Exploration Society. (1989)

Bibby, G., *Four Thousand Years Ago: a panorama of life in the second millenium BC*, Collins, London. (1962)

Biggs, R., 'Inscriptions from Tell Abu Salabikh', *Oriental Institute Publications 99*, University of Chicago, Chicago, Illinois. (1974)

— 'My Career in Assyriology and Near Eastern Archeology', *Oriental Institute*, University of Chicago, Chicago, Illinois. (2005)

Bilkadi, Zayn, 'Bitumen – A History', *Saudi Aramco World*, (1984)

Boëda, E., Connan, J. and Muhesen, S., 'Bitumen as hafting on Middle Paleolithic artifacts from the El Kown Basin, Syria', *Neanderthals and modern humans in Western Asia'*, pp. 181-204 (eds. T. Akazawa, K. Aoki and O. Bar-Yosef), Plenum Press, New York. (1998)

Braslavski, Y., 'The Earthquake and the stoppage of the Jordan River in 1546', *Zion* 3:323-336 (in Hebrew). (1938)

Braslavy, Y., *Around the Dead Sea*, Hakibbutz Hameuhad, Tel Aviv. (1956)

British Broadcasting Corporation, *Ancient Apocalypses: Sodom and Gomorrah.* (2001)

Brugsch-Bey, H., *A History of Egypt under the Pharaohs*, Murray, London. (First printed in 1879, reprinted 1902)

Budge, E.A.T.W., *Babylonian Life and History*, Scott, London. (1884)

— *The Mummy; Chapters on Egyptian Funereal Archaeology*, Cambridge University Press, Cambridge. (1893)

Bunbury, E.H., *A History of Ancient Geography*, Dover, New York. (1959)

Cassius Dio, *Historia Romana*, Harvard University Press, Loeb Classical Library Volume VIII. (1925)

Champkin, J., 'Hunting for hell on earth', *Weekend Mail.* (July 14, 2001)

Clapp, G.F., 'Geology and bitumens of the Dead Sea area, Palestine and Transjordan', *American Association of Petroleum Geologists Bulletin*, vol. 20, pp. 881-909. (1936)

Close, U., and McCormick, E., 'Where the Mountains Walked', *The National Geographic Magazine*, vol. XLI, no. 5, Washington. (1922)

Closson, D., Karaki, N.A., and al-Fugha, H., 'Subsidence hazards in the Dead Sea area, evidence from radar interferometry', *Eighth International Conference of Jordanian Geologists*, Jordan University, Jordan. (April 6-7, 2004)

Connan, J., 'Use and trade of bitumen in antiquity and prehistory: molecular archaeology reveals secrets of past civilizations', *Philosophical Transactions of the Royal Society Biological Sciences* B354: 33-50. (1999)

Connan, J. and Deschesne, O., 'Archaeological bitumen identification, origins and uses of an ancient near eastern material', *Materials Research Society Symposium Proceedings*, vol. 267, pp. 683-720, Pittsburgh. (1995)

Connan, J., Nissenbaum, A., and Dessort, D., 'Molecular Archaeology: Export of Dead Sea Asphalt to Canaan and Egypt in the Chalcolithic-Early Bronze Age (4th-3rd millenium BC)', *Geochimica et Cosmochimica Acta*, vol. 56, pp. 2743-2759, Elsevier, Amsterdam. (1992)

Coogan, M.D., 'Numeira 1981', *Bulletin of the American Schools of Oriental Research*, vol. 255; 75-81. (1984)

Crawford, C.B., 'Cohesion in an undisturbed sensitive clay', *Geotechnique*, London. (1963)

Davidovits, J., *Alchemy and the Pyramids*, Geopolymer Institute, Saint Quentin, France. (1983)

— *Geopolymer Chemistry and Applications*, Geopolymer Institute, Saint Quentin, France. (2008)

Davidovits, J. and Morris, M., *The Pyramids: An Enigma Solved*, Dorset Press, New York. (1988)

Dawson, G.M., 'Remarkable landslip at Portneuf County, Quebec', *Bulletin of the Geological Society of America*, vol. 10. (1899)

Diodorus Siculus, *Bibliotheca Historica*, in *The Bibliotheca historica of Diodorus Siculus*, Oxford University Press for the Early Text English Society, Oxford. (1957). Translated by John Skelton.

Dioscorides, *De Materia Medica*, Olms-Weidmann, Hildesheim. (2005) Translated by Lily Y. Beck.

Donahue, J., 'Geology. Preliminary Report of the 1979 Expedition of the Dead Sea Plain Expedition, Jordan' (eds. Rast, W.E. and Schaub, R.T.), *Bulletin of the American Schools of Oriental Research* 240: 47-52. (1980)

— 'Geologic Investigations at Early Bronze Sites. An Interim Report of the 1977 Season: The Southeastern Dead Sea Plain Expedition', (eds. Rast, W.E. and Schaub, R.T.), *Annual of the American Schools of Oriental Research* 46: 137-54. (1981)

— 'Hydrologic and Topographic Change During and After Early Bronze Occupation at Bab edh-Dhra', *Studies in the History and Archaeology of Jordan*, (ed. A. Hadidi), Amman: Department of Antiquities, 2: 131-40. (1985)

Druckman, Y., Magaritz, M. and Sneh, A., 'The shrinking of Lake Lisan as reflected by the diagenesis of its marginal oolitic deposits', *Israel Journal of Earth Sciences*, 36:101-106. (1987)

Eiby, G.A., *Earthquakes*, Frederick Muller, London. (1967)

Elazari-Volcani, B., 'Bacteria in the bottom sediments of the Dead Sea', *Nature*, 152: p. 274. (1943a)

— 'A Dimastigamoeba in the bed of the Dead Sea', *Nature*, 152: p. 301. (1943b)

Emery, K.O., 'A reconnaissance study of the floor of the Dead Sea', *Search for Sodom and Gomorrah*, CAM Press, Kansas City, Missouri, (ed. Baney, R.E.), pp. 305-320. (1962)

Enzel, Y., Kadan, K. and Eyal, Y., 'Holocene Earthquakes Inferred from a Fan-Delta Sequence in the Dead Sea Graben', *Quaternary Research* 53:34-48. (2000)

Esse, D.L., 'Secondary state formation and collapse in Early Bronze age, Palestine', *L'urbanisation de la Palestine a l'age du Bronze Ancien: Actes du colloque d'Emmaus, 20-24 Octobre 1986* (ed. P.R. de Miroschedji), British Archaeological Reports, International Series 527. (1989)

Eugster, H.P. and Hardie, L.A., 'Saline Lakes', *Lakes: Chemistry, Geology, Physics*, Springer, Berlin. (1978)

Forbes R.J., *Bitumen and Petroleum in Antiquity*, E.J. Brill, Leiden. (1936)

— *Studies in Ancient Technology*, vols. 1-9, E.J. Brill, Leiden. (1964/72)

Forbes, R.J., and Dijksterhuis, E.J., *A History of Science and Technology (vol. 1)*, Penguin Books, London. (1963)

Frain, I., *L'inimitable*, Fayard, Paris. (1998)

Freedman, D.N., Response to a letter by Carl McIntire (25 Nov 1976), *The Christian Beacon*, vol. 25. (1977)

Friedman, M., 'Decades of work uncovers changes in Old Testament', *The Guardian*, Charlottetown. (August 13, 2011)

Frostick, L.E., and Reid, I., 'Climatic versus tectonic controls of fan sequences: lessons from the Dead Sea, Israel', *Journal of the Geological Society*, 146(3):527-538. (1989)

Frumkin, A., 'The Holocene history of Dead Sea levels', in *The Dead Sea, the lake and its setting*, (eds. Niemi, T.M., Ben-Avraham, Z., and Gat, J.), Oxford University Press, Oxford. (1997)

Frumkin, A., 'Formation and dating of a salt pillar in Mount Sedom diapir, Israel', *Bulletin of the Geological Society of America*, vol. 121(1-2): 286-293. (2009)

Frumkin, A., Magaritz, M., Carmi, I. and Zak, I., 'The Holocene climatic record of the salt caves of Mount Sedom, Israel', *The Holocene* v1: 191-200. (1991)

Frumkin, A., and Ford, D.C., 'Rapid entrenchment of stream profiles in the salt caves of Mount Sedom, Israel', *Earth Surface Processes and Landforms*, 30:139-152, Wiley-Blackwell. (1991)

Gardosh, M., Kashai, E., Shalhov, S., Shulman, H. and Tannenbaum, E., 'Hydrocarbon exploration in the southern Dead Sea area', *The Dead Sea: the lake and its settings,* (eds. Niemi, T.M., Ben-Avraham, Z. and Gat, J.), Oxford University Press, New York. (1997)

Garstang, J. 'The fate of Jericho as revealed by the spade', *London News*. (16 Dec 1933)

Gary, M., McAfee, R. and Wolf, C.L. (eds.), *Glossary of Geology*, American Geological Institute, Washington. (1974)

Geological Survey of Israel, 2000, *Bathymetric Map of the Dead Sea*, GSI Report GSI/16/2000 (Bathymetry after surveys by Lt. W.F. Lynch, David Neev, and John Hall, first published 1975).

Glueck, N., 'An aerial reconnaissance in Southern Transjordan', *Bulletin of the American School of Oriental Research*, 67: 19-26. (1937)

Gonen, R., 'Shepherds, farmers, and coppersmiths – the Chalcolithic Period', in *The Archaeology of Ancient Israel in the Biblical Period* (in Hebrew, ed. A. Ben-Tor) pp. 119-183, Open University of Israel, Tel Aviv. (1989)

Gophna, R., 'The Intermediate Bronze Age', in *The Archaeology of Ancient Israel* (ed. A.Ben-Tor), Yale University Press, New Haven. (1992)

Hadas-Lebel, M., *Flavius Josephus: Eyewitness to Rome's first century conquest of Judea*, Macmillan, New York. (1993)

Haigh, S.K. and Madabhushi, S.P.G., 'Dynamic centrifuge modelling of the destruction of Sodom and Gomorrah', *Proceedings of the International Conference on Physical Modelling in Geotechnics: ICPMG '02*, Guo and Popescu (eds.), Balkema, Rotterdam, pp. 507-511. (2002)

Har-El, M., *The Judean Desert and the Dead Sea* (in Hebrew), Am-Oved Publishers, Jerusalem. (1971)

Harland, J.P., 'Sodom and Gomorrah – The location of the Cities of the Plain', *The Biblical Archaeologist*, 5:17-32. (1942)

Harris, G.M. and Beardow, A.P., 'The destruction of Sodom and Gomorrah: a geotechnical perspective', *Quarterly Journal of Engineering Geology*, 28:349-362, London. (1995)

Harris, G.M., and Partridge, T.C., 'Engineering properties of a delta deposit of Lake Malawi', *Seventh regional Conference for Africa on Soil Mechanics and Foundation Engineering*, Accra, Ghana. (1980)

Hartel, M., Hazan, N., Lev, L. and Stein, M., 'The AD 749 earthquake, Dead Sea transform', *Geology* 31(8):665-668. (2003)

van Hattem, W.C., 'Once again: Sodom and Gomorrah', *The Biblical Archaeologist*, Boston, Massachusetts. (1981)

Husein, A., Al-Homoud, A.S. and Liang, R.Y., 'Seismic hazard mapping of Jordan', *Quarterly Journal of Engineering Geology*, 28: 75-81, London. (1995)

Irby, C.L., and Mangles, J., *Travels in Egypt and Nubia, Syria and Asia Minor during the years 1817 and 1818*, T. White, London: pp. 453-55. (1823)

Jassim, S.Z., Raiswell, R. and Bottrell, S.H., 'Genesis of the Middle Miocene stratabound sulfur deposits of northern Iraq', *Journal of the Geological Society*, vol. 156, pp. 25-39, London. (1999)

Josephus, F., *The Wars of the Jews*, digireads.com Publishing. (Translated by William Whiston.)

Keller, W., *The Bible as History: Archaeology Confirms the Book of Books*, Hodder and Stoughton, London. (1956)

Kelly, G., 'Ammianus and the Great Tsunami', *Journal of Roman Studies*, Society for the Promotion of Roman Studies, 94:141-167, London. Also see Thucydides *Peloponnesian War*, Book 1.22. (2004)

Ken-Tor, R., Agnon, A., Enzel, Y., Marco, S., Negendank, J.F.W. and Stein, M., 'High resolution geological record of historic earthquakes in the Dead Sea basin', *Journal of Geophysics Research*, 106:2221-2234. (2001)

Kenyon, K.M., *Archaeology in the Holy Land*, Benn, London. (1979)

Klein, C., *On the Fluctuations of the Level of the Dead Sea since the beginning of the 19th Century*, Hydrological Paper 7, State of Israel Hydrological Service. (1961)

— *On the Fluctuations of the Level of the Dead Sea since the beginning of the 19th Century*, Hydrological Paper 7, State of Israel Hydrological Service (revised edition). (1965)

— Morphological evidence of lake level changes, western shore of the Dead Sea, *Israel Journal of Earth Sciences*, 31, pp. 67-94. (1982)

Kuhn, Oliver, 'Odyssey of Oil', *CSEG Recorder* (a publication of the Canadian Society of Exploration Geophysicists), 25, pp. 31-38. (2000)

Kuhrt, A., *The Ancient Near East c. 3000-330 BC*, Routledge, London. (1995)

Kuribayashi, E. and Tatsuoka, F., 'Brief review of soil liquefaction during earthquakes in Japan', *Soils and Foundations* 15:81-92, Japanese Geotechnical Society, Tokyo (Also in *Wind and Seismic Effects*, Proceedings of the 7th Joint Panel Conference of the U.S.-Japan Cooperative Program in Natural Resources (May 20-23, 1975) Tokyo.) (1975)

Kyle, M.G., *Explorations at Sodom: The story of ancient Sodom in the light of modern research*, Robert Scott, London. (1928) Republished by Arno Press, New York, 1977.

Lamb, H., *The Crusades: Iron Men and Saints*, Doubleday, New York. (1953)

Layard, A.N., *The Monuments of Nineveh*, John Murray, London. (1849)

Lein, A.Y., Samarkin, V.A., Zyakun, A.M. and Matrosov, A.G., 'On the genesis of native sulfur in the Kerch peninsula', *Geochemistry International*, vol. 12, pp. 15-25, Moscow. (1975)

Lingojsky, Y., 'Hitting the gas jackpot in the Mediterranean', *International Herald Tribune*, Israel, (September 26, 2000)

Lynch, W.F., *Narrative of the United States Expedition to the River Jordan and the Dead Sea*, Richard Bentley, London. (1849)

Marco, S., Stein, M., Agno, A. and Ron, H., 'Long term earthquake clustering: a 50,000 year paleoseismic record in the Dead Sea graben', *Journal of Geophysics Research*, 101(B3):6179-6192. (1996)

Marinatos, S.N., 'Helice Submerged Town of Classical Greece', *Archaeology*, vol. 13:3, New York. (1960)

Maundeville, J., *The Voyage and Travels of Sir John Maundeville, Kt.*, Cassell, London. (1836)

Mazar, B., Dothan, T. and Dunayevsky, I., 'En Gedi, the first and second seasons of excavations 1961-1962', *Atiqot* 5: 1-100. (1966)

McCalpin, J., *Paleoseismology* (2nd edition), Academic Press, London. (2009)

McCreery, D.W., 'Flotation of the Bab edh-Dhra and Numeira Plant Remains', in *The Southeastern Dead Sea Plain Expedition: An Interim Report of the* 1977 *Season* (ed. W.E. Rast and R.T. Schaub), pp. 165-69, *Annual of the American Schools of Oriental Research* 46, ed. J.A. Calloway. (1981)

Mitchell, T.C., *The Bible in the British Museum*, British Museum Press, London. (1988)

Montague, E.P., *Narrative of the Late Expedition (1847-48) to the Dead Sea, from a diary of one of the party.* (1849)

Musil, A., *Arabia Petraea, II Edom*, A. Hölder, Vienna. (1907)

Neev, D., *The Dead Sea*, Geological Survey of Israel, Report Q/2/64, Jerusalem. (1964)

Neev, D. and Emery, K.O., *The Dead Sea – Depositional Processes and Environments of Evaporites*, Geological Survey Bulletin 41, State of Israel. (1967)

— *The Destruction of Sodom, Gomorrah and Jericho: Geological, Climatological and Archaeological Background*, Oxford University Press, Oxford. (1995)

Niemi, T.M., Ben-Avraham, Z. and Gat, J. (eds.), *The Dead Sea: the lake and its setting*, Oxford University Press, New York. (1997)

Nissenbaum, A., 'Dead Sea Asphalts: Historical Aspects', *American Association of Petroleum Geologists Bulletin*, vol. 62: 837-844. (1978)

— 'Sailing routes in the Dead Sea', *Weitzman Foundation Journal for Natural Sciences and Technology*, vol. 21, no. 10, pp. 4-6. (1990)

— 'Molecular Archaeology: Organic Geochemistry of Egyptian Mummies', *Journal of Archaeological Science*, vol. 19. (1992)

Nissenbaum A., Aizenshtat, Z. and Goldberg, M., 'The floating asphalt blocks of the Dead Sea', *Advances in Organic Geochemistry*. (eds. Douglas and Maxwell), Pergammon Press, Oxford. (1979)

Nissenbaum, A., Serbon, A., Amran. R., and Ilam, O., 'Dead Sea asphalt from the excavations in Tel Arad and Small Tel Malhata', *Paléorient*, vol. 10 (pp. 157-161), Paris. (1984)

Nur, A., and MacAskill, C., *The Walls Came Tumbling Down – Earthquakes in the Holy Land*, Stanford University, Department of Geophysics (VHS Stereo). (1990)

Nur, A. and Ron, H., 'Earthquake! Inspiration for Armageddon', *Biblical Archaeological Review*, 23:04, Washington, D.C. (1997)

Partington, J.R., *A History of Greek Fire and Gunpowder*, Heffer, Cambridge. (1960)

Peters, K.E., Walters, C.C. and Moldowan, J.M., *The Biomarker Guide*, Cambridge University Press, Cambridge. (2005)

Petre, J., 'Sodom and Gomorrah found at bottom of Dead Sea', *The Sunday Telegraph*, London. (March 26, 2000)

Porat, N., *Composition of pottery – Application of the study of the interrelations between Canaan and Egypt during the third millennium BC*, Jerusalem (in Hebrew). Unpublished PhD thesis, Hebrew University. (1989)

Pliny, *Historiae Naturalis. The Elder Pliny's Chapters on Chemical Subjects,* London. (1929, 1932) Translated by Bailey.

von Rad, G., *Genesis*, Old Testament Library, Westminster Press, Philadelphia. (1971)

Rappoport, A. S., *Ancient Israel: Myths and Legends*, Gresham, London. (1928). Recently reprinted by Studio Editions, London 1995.

Rast, W.E., 'Settlement at Numeira' (pp. 35-44) in *The Southeastern Dead Sea Plain Expedition: An Interim Report of the* 1977 *Season*, (eds. W.E. Rast and R.T. Schaub), *Annual of the American Schools of Oriental Research* 46, (ed. J.A. Calloway). (1981)

— 'Bab edh-Dra'a and the origin of the Sodom saga', in *Archaeology and the Biblical Interpretation* (eds. D.L. Pardee, L. Toombs and G. Johnson), John Knox Press, Atlanta. (1987)

— 'Bab edh-Dra', in *Anchor Dictionary of the Bible* (ed. D.N. Freeman), 5:895-6, Doubleday, New York. (1993)

Rast, W.E., and Schaub, R.T., 'Survey of the Southeastern Plain of the Dead Sea, 1973', *Annual of the Department of Antiquities of Jordan*, 19:5-54, 175-85, Hashemite Kingdom of Jordan, Department of Antiquities. (1974)

— 'Preliminary Report of the 1979 Expedition to the Dead Sea Plain, Jordan', *Bulletin of the American Schools of Oriental Research* 240: 21-61. (1980)

Reifenberg A., *The Struggle of the Desert and the Sown. Rise and fall of agriculture in the Levant,* Dept. of the Jewish Agency, Jerusalem. (1955)

Rizkana, I. and Seeher, J., – Maadi III., 'The non-lithic small finds and the structural remains of the pre-dynastic settlement', *Mitteilungen des Deutschen Instituts für Ägyptische Altertumskunde in Kairo*, pp. 71-72. (1989)

Roux, G., *Ancient Iraq*, George Allen and Unwin, London. (1964)

Sauer, S.P., 'Recent and projected changes in Dead Sea level and effects on mineral production from the sea', *U.S. Geological Survey Open-file Report 78-176*, Reston, Virginia. (1978)

Schaub, R.T., 'Feifa and Khazir' in Archaeology in Jordan, *American Journal of Archaeology*, 95:254-255. (1991)

— 'Safi' in *Anchor Dictionary of the Bible* (ed. D.N.Freeman), 5:895-96, Doubleday, New York. (1992)

— Bab edh-Drha in *The New Encyclopedia of Archaeological Excavations in the Holy Land*, (ed. E.Stern), Simon and Schuster, New York. (1993)

Schmidt, J.F., *Studien über Erdbeben*, A. Georgi, Leipzig. (1875)

Seed, H.B., *Landslides during Earthquakes due to Soil Liquefaction*, Fourth Terzaghi Lecture, in *Terzaghi Lectures*, 1963-1972, pp. 193-261, American Society of Civil Engineers, New York. (1967)

Seed, H.B. and Idriss, I.M., *Ground Motions and Soil Liquefaction during Earthquakes*, Earthquake Engineering Research Institute, Berkely, California. (1982)

Shapira, A., 'On the seismicity of the Dead Sea basin', *The Dead Sea: the lake and its settings*, (eds. Niemi, T.M., Ben-Avraham, Z. and Gat, J.) Oxford University Press, New York. (1997)

Shea, W. H., 'Ebla Reveals Her Secrets', *Ministry*, vol. 52, 28-30. (1979)

Silberman, N.A., 'Sodom and Gomorrah Update', *Archaeology Newsbriefs*, vol. 49, no. 4. (1996)

Stiller, M., Carmi, I. and Kaufman, A., 'Organic and inorganic ^{14}C concentrations in the sediments of Lake Kinnaret and the Dead Sea (Israel) and the factors which control them', *Chemical Geology*, vol. 73:63-78, Amsterdam. (1988)

Strabo, *Geographica*. (ed. Jones), Loeb Classics Library. See also *The Geography of Strabo*, William Heinemann, London. (1930). Translated from the Greek by H.L.Jones.

Tactitus, *Aineias Tacitus: How to Survive under Siege*, Bristol Classic Press, London. (2002). Translated by D. Whitehead.

Torfstein, A., Gavrieli, L., Katz, A., Kolodny, Y. and Stein, M., 'Gypsum as a monitor of the paleolimnological-hydrological conditions in Lake Lisan and the Dead Sea', *Geochimica et Cosmochimica Acta*, vol. 72, pp. 2491-2509, Elsevier, Amsterdam. (2008)

Trifonov, V.G., 'The Bible and geology: destruction of Sodom and Gomorrah', in *Myth and Geology*, (eds. L. Piccardi, and W.B. Masse), Geological Society of London, Special Publications, vol. 273, pp. 133-142. (2007)

Tubb, Jonathan, *Canaanites: Peoples of the Past*, British Museum Press, London. (1998)

Turcotte, T. and Arieh, E., 'Catalog of earthquakes in and around Israel', *Preliminary safety analysis report*, Israel Electric Corporation, Tel Aviv, Israel. (1988)

Uren, L.C., *Petroleum Production Engineering: Oil Field Development*, McGraw-Hill, New York. (1946)

Vawter, B., *On Genesis – A New Reading*, Doubleday, New York. (1977)

Vitruvius, *De Architectura*, in *The Architecture of M. Vitruvius Pollio*, translated from the Latin by W. Newton, James Newton, London. (1791)

Wang, W., *Some Findings in Soil Liquefaction*, Water Conservancy and Hydroelectric Power Division, Scientific Research Institute, Beijing. (1979)

Watson, C.M., 'The stoppage of the River Jordan in AD 1267', *Quarterly Journal of the Palestine Exploration Fund*, pp. 253-261. (1895)

Weinstein, J., 'Radiocarbon dating in the Southern Levant', *Radiocarbon*, vol. 26, no. 3, pp. 297-366. (1984)

Wood, B.G., 'The Discovery of the Sin Cities of Sodom and Gomorrah', *Bible and Spade*, Akron, Pennsylvania. (1999)

Woolley, C.L., *The Sumerians*, Clarendon, Oxford. (1928)

—*The Ziggurat and its Surroundings* (UE,V), London. (1939)

Wright, G.E., 'The Archaeology of Palestine, in the Bible and the Ancient Near East', in *Essays in Honor of W.F. Albright* (ed. G.E. Wright), Routledge and Kegan, London. (1961)

Wright, M., 'Contacts between Egypt and Syro-Palestine during the Protodynastic Period', *Biblical Archaeology*, vol. 48, pp. 240-253. (1985)

Yechieli, Y., Magaritz, M., Levy, Y., Weber, U., Kafri, U., Wölfli, W. and Bonani, G., 'Late quaternary geological history of the Dead Sea area, Israel', *Quaternary Research* 39:59-67, Elsevier, Amsterdam. (1993)

Youd, T.L., and Hoose, S.N., 'Liquefaction susceptibility and geologic setting', *Proceedings of the 6th World Conference on Earthquake Engineering*, pp. 2189-2194. (1977)

Zak, I., *The geology of Mount Sedom* (PhD thesis), The Hebrew University, Jerusalem. (1967)

Zeng, Z., Chen, C-T. A., Yin, X., Zhang, X., Wang, X., Zhang, G., Wang, X. and Chen, D., 'Origin of native sulfur ball from the Kueishantao hydrothermal field offshore northeast Taiwan', *Journal of Asian Earth Sciences*, vol. 40, pp. 661-671, Elsevier, Amsterdam. (2011)

Zori, N., 'An archaeological survey of the Beth-Shean Valley' (in Hebrew), in *The Beth-Shean Valley: A Colloquium of Papers*, The 17th Archaeological Convention, Israel Exploration Society, Jerusalem, pp. 135-98. (1962)

Glossary

This glossary has been assembled to assist the reader in understanding some of the more unfamiliar terminology in the text. It is not exhaustive, and reference should be made to more comprehensive works for further details.

Alluvium
Sediments deposited from rock particle detritus carried in stream or river flow.
Anhydrite
A relatively soft, colourless to whitish-grey or lilac-coloured evaporite. Chemical formula $CaSO_4$. Found frequently in association with gypsum and salt in sedimentary rock formations, all of which are derived from the evaporation of sea water.
Apigravity
A measure of how heavy or light a petroleum product is with respect to water. The American Petroleum Institute assigns a gravity of 10 to water, with higher values indicating lighter oils that float, and lower values to oils that sink.
Aragonite
A mineral form of calcium carbonate which, though chemically identical, possesses a different form of crystallisation, and is of higher specific gravity.
Asphalt
See Bitumen.

Bathymetry
Contours of undersea land forms.
Bioepigenesis
The growth of living organisms by means of sequential cell development.
Biogenesis
The creation of new living organisms, including those created from non-living matter.
Bitumen
Unrefined, naturally occurring material. See Chapter 4 (sub–section 'Etymology and Nomenclature') for a detailed description.

Carbon Dating
A method of dating organic material by determining the remnant amount of carbon-14 present. After the death of an organism, decay of the carbon-14 ensues. By measuring the emission of radiation it is possible to determine the approximate period of time that has elapsed since the death of the organism. See Appendix I.
Carnallite
A double salt of potassium and magnesium chlorides ($KCl.MgCl_2.6H_2O$).
Cenomanian
The earliest age of the Late Cretaceous (100-94 million years ago). It underlies the Turonian (94-89 million years ago).

Centrifuge Testing
A laboratory technique for testing scale models of important facilities in various applications of geotechnical engineering, where forces produced by centrifugal acceleration simulate those of gravity. A.N. Schofield of Cambridge University played a key role in the development of modern methods.
Chalcolithic
Also known as the Copper Age, a phase of the Bronze Age before metal-workers discovered that adding tin to copper produced a harder bronze. The period is roughly estimated as being from 5300 BC to 1700 BC, but highly variable depending on region.
Chert
A hard fine-grained siliceous material of sedimentary origin, being deposited from sea water under favourable conditions. Often known as flint where it occurs in nodular form.
Chromatogram
See Gas Chromatography
Conglomerate
A rock composed of cemented sand, gravel, and boulders.
Cretaceous
A period of generally warm climate conducive to the creation of life 145 to 66 million years ago. The Cretaceous is divided into Lower, Middle, and Upper.
Cyclic Shear Stress
The stress imposed on a material by external vibratory forces, such as from earthquake or other external shock loadings. When the cyclic shear stresses exceed the shear strength failure occurs.

Diapir
An intrusion into overlying rocks from below. When evaporites, such as salt, are overlain by thick deposits of sediments they become ductile, and the magnitude of the pressures involved can inject the salt into the overlying sediments at points of weakness.

Escarpment
An abrupt change in topography rising to great height, often formed by landslide or faulting caused by earthquake.
Evaporite
A sedimentary rock formed by the evaporation of sea water, e.g. salt, gypsum, anhydrite, and forms of limestone.

Gas Chromatography
A laboratory technique that volatilises and separates molecular constituents of an organic mixture for identification purposes without altering their molecular structure. The individual components are identified on a chart, known as a chromatogram.
Geochronology
The branch of geology concerned with the dating of rock formations.
Geomorphology
The study of those physical features of Earth's crust, as a consequence of geological agencies throughout its history.
Graben
A depressed block of land bounded by parallel fault systems. The valley, or trench, so formed often possesses steep scarps on each side. (From *graben*, the German for trench)

Gypsum
A soft, colourless to white-grey rock. Chemical formula $CaSO_4.2H_2O$). Gypsum often forms from anhydrite by hydration under low pressure, during which it expands creating fractures. Gypsum can dehydrate under pressure, whereby it is transformed back into anhydrite.

Holocene
A geological epoch which began with the end of the Pleistocene 11,700 years BP), and continues to the present.

Intensity (Modified Mercalli Scale)
A qualitative measure of the damage caused by an earthquake. See Appendix II.

Limestone
A widely distributed sedimentary rock having the chemical formula $CaCO_3$. The degree of hardness can vary considerably.

Liquefaction
The ability of a saturated soil to flow 'like a liquid', after being subjected to external forces or shock loadings incapable of being withstood by the soil structure. Loose sands, silts, and certain clays are vulnerable to liquefaction.

Liquid Limit
A geotechnical term reflecting the water content of a clay at which its behavioural characteristics alter from plastic to liquid.

Magnitude (Richter Scale)
A quantitative measure of the seismic energy released by earthquake. See Appendix II.

Marl
Clays or silts of sedimentary origin containing a significant percentage of calcium carbonate and associated minerals, e.g. gypsum or anhydrite.

Metamorphic Rocks
Rocks of considerably different properties from their parents as a result of being subjected to intense pressure and/or heat.

Mesozoic
An interval of geologic time from about 252 to 66 million years ago. Also known as the Age of Reptiles.

Micrite
A fine-grained limestone of calcareous composition, often unconsolidated, e.g. in the form of muds deposited in an evaporitic environment.

Miocene
An epoch lasting from 23 to 5.3 million years before present, during which much life as we know it began to flourish.

Morphology
The science and study of landforms resulting from natural agencies.

Natron
Sodium carbonate, or soda ($Na_2CO_3.10H_2O$)

Neolithic
A period of human development commencing about 10,200 BC and ending between 4500 BC and 2000 BC depending on location. Also known as the New Stone Age.

Nubian Sandstone
A ferruginous sandstone widespread throughout the Middle East, and extending well

into North Africa. Of continental origin it is of uncertain age in many locations due to the absence of fossils. Towards the southeastern area of the Dead Sea it is of some 600 metres thick, and considered Cretaceous.

Paleozoic
A geologic era spanning the period 541 to 253 million years before present. (From the Greek meaning 'ancient life').
Paleontology
The branch of science concerned with the study of fossil remains of animals and plants.
Palynology
The study of pollen grains and spores, especially those connected with archaeological remains.
Plastic Limit
A geotechnical term reflecting the water content of a clay at which its behavioural characteristics alter from plastic to non-plastic.
Playa
An area of flat, dried-up land, often in a desert basin.
Pleistocene
A geologic epoch lasting from 2.5 million years ago to 11,700 years before the present, during which major repeated glaciations occurred.
Plethron
A measure of length equal to a hundred Greek feet, or 40 paces, i.e. about 30 metres.

Quaternary
A geologic period spanning from 2.6 million years ago to the present, and characterised by numerous repeated glaciations and the appearance of humans.

Radiocarbon Dating
See Carbon Dating.

Sedimentary Rocks
Rocks laid down in the form of sediments as a consequence of erosion of other rock types. Common sedimentary rock types are shale, mudstone, siltstone and sandstone. Limestone, gypsum, anhydrite and coal, are also considered to be within this class.
Sensitivity
The ratio of the strength of a soil in its natural intact state, to the strength when its internal fabric has been destroyed by remoulding.
Shear Strength
The strength of a material at which yield or failure occurs when subject to shear stress.
Shear Stress
The stress exerted by a force vector coplanar with a material's cross-section. When shear stress exceeds shear strength substantial yield or failure occurs.
Sinkhole
A cavity, often of considerable size, caused by solution or erosion of the underlying soil/rock leading to collapse of the surface strata.
Slate
A metamorphic rock produced from shale.
Stadium
A Roman measure of length equal to about 185 metres.
Stratum
A layer of soil or rock that possesses common properties and characteristics.

Strike-Slip Movement
Rupture due to earthquake whereby the two sides move horizontally with no relative vertical displacement.
Sylvinite
Potassium chloride (KCl).

Tectonic Plates
Crustal plates of Earth's surface, often in movement with one another to cause earthquakes.
Triassic
An interval of geologic time of between 250 to 200 million years ago.
Tsunami
A wave, or series of waves, often destructive in size and effect, created by soil or rock movement resulting from earthquake or landslide displacing a body of water.
Turonian
See Cenomanian.

Void Ratio
A term used in materials science reflecting the ratio of pore volume, whether occupied by air or liquid (usually water), to the volume of solids.
Volcanic Rocks
Igneous rock types that have originated from the cooling of lavas ejected from volcanoes, e.g. basalt. The term also includes consolidated ashes and ejecta. Volcanic rocks are invariably fine-grained as a result of rapid cooling of the lava. There is little or no crystallinity apparent to the naked eye.

Appendix I.
A Note on Radio-Carbon Dating

Radiocarbon dating of organic materials was developed by Walter Libby and his colleagues at the University of Chicago in 1949. The first demonstration of the accuracy of this new breakthrough was vividly demonstrated when a fragment of wood removed from an ancient Egyptian royal barge was accurately dated, the date of which was known from historical documents. Libby was awarded the Nobel Prize for chemistry in 1960.

What follows is a simplified explanation of the basis behind radiocarbon dating theory, with a brief discussion of some of the flaws which can lead to erroneous conclusions. This note is no substitute for the many excellent texts available, which treat the subject in considerably greater depth, and which interested readers may wish to seek out for themselves.

Earth is under continual bombardment by cosmic rays from outer space. It is assumed that the rate of bombardment is constant (though this is far from being the case). Cosmic rays are composed of various sub-atomic particles, and in the process of interaction high energy neutrons are produced which impact on atmospheric nitrogen. Protons are knocked out of the nitrogen nuclei, the end result being the formation of a radio-active isotope of carbon which possesses two extra neutrons in its nucleus, i.e. it becomes ^{14}C (carbon-14). The interchange may be perceived as:

$$n + {}^{14}_{7}N^{\,1+} \rightarrow 6\ {}^{14}_{6}C + p$$

Non-radio-active isotopes of carbon are ^{13}C and ^{12}C, mostly the latter, and, as the radio-active carbon (^{14}C) decays it progressively mutates into the stable form ^{12}C (carbon-12). The half-life of carbon of this decay process is 5730 ± 40 years.

It is assumed the carbon-14 permeates the atmosphere uniformly, though it will be appreciated that the denser the atmosphere becomes at lower altitudes the rate at which carbon-14 is produced will diminish. At very low elevations, such as at the Dead Sea, carbon-14 production is likely to be lower than many other locations on the planet. Also, it has been shown that there are latitudinal and continental variations in its atmospheric concentration. The greatest production of carbon-14 is considered to be in the zone from nine to fifteen kilometres above Earth's surface. The ratio of carbon-14 to non-radioactive carbon is approximately one part per trillion. Thus the concentration of carbon-14 in the atmosphere is exceedingly small.

Living organisms absorb carbon-14 by photo-synthesis of flora, and by ingestion if they are fauna, and this intake accumulates until death. Decay of carbon-14 commences following its assimilation, the decay process capable of being measured by a number of laboratory techniques, all of which are based on proven science and technology. It serves no purpose to discuss here the suitability of specific techniques for specific applications, except to note that some are better than others (and more expensive) – ordinary mass spectrometry, for instance, sometimes fails to differentiate between radioactive carbon-14 and ordinary nitrogen! These laboratory techniques can be used to determine ages of carbonaceous materials up to about 60,000 years, which are generally reported in terms of 'years BP', i.e. years before the present. The benchmark year for this is 1950, as this was the year when it became apparent atmospheric testing of nuclear weapons had contributed significant amounts of radiation to the atmosphere, thus rendering useless later applications of the technique.

A number of anomalous situations can arise by thoughtless application of the results from radiocarbon dating. The case of the stick of wood discovered in the burial chamber at Bab edh-Drha has been noted already, as well as the controversy over the Shroud of Turin. Primordial carbon dioxide emitted by volcanoes, or fossil carbon dioxide associated with the consumption of fossil fuels, is likely to have a significant impact on the carbon-14 intake of both flora and fauna. Thus, age determinations from testing may appear appreciably older. To demonstrate this point two case histories may be mentioned. Firstly, some living crustaceans were crushed and their remains tested; these were determined to be about 2500 years old. The creatures had formed their shells from calcium carbonate extracted from the sea. Secondly, some coconut fibre from the West Indies was tested, and determined to have an age dating it to the twelfth century. The result was absurd from knowledge of the facts, a date of 'after 1720' being subsequently assigned, as this was the year when the Mexican volcano, Popocatepetl, exploded violently unleashing massive amounts of primordial carbon dioxide into the atmosphere.

Another difficulty in assessing the impact of fossil carbon on radiocarbon test results was encountered in attempting to determine the age of the Sedom Clay in the vicinity of the Dead Sea Works chemical plant at S'dom. Though some fragments of wood were recovered in one of the borings they were considered too small and inadequate for testing. Various alternatives to radiocarbon dating were considered, but eliminated for a number of reasons. In the end an attempt was made to determine the age of the carbon in the carbonate, the clay being micritic, i.e. composed essentially of calcium carbonate. The date that arose was 23,600 ± 300 years BP. Because of the certainty of the inclusion of older carbon, and its effect upon the result, this date was perceived as representing an ultimate 'upper-bound', the true age being younger by an unknown number of years. Taking all other available information into account the Sedom Clay was later assessed as having been deposited between 17,000 and 20,000 years BP.

Various contaminants may seriously affect radiocarbon dating leading to younger dates being determined. Such contaminants can include the presence of organic residues of recent formation, which might include air-borne bacteria, mildew, spores of fungi and pollen, as well as pigments previously applied. In addition there are oils and greases which may inadvertently cause contamination of the surface. These may have an air-borne origin, or result from human handling as possible with the Shroud of Turin.

There is, also, the well-known 'ring effect' when dating specimens of wood. Assume a tree possesses an age of five hundred years before being hewn. There is, therefore, a five hundred year age difference between the older heartwood at its centre, and the younger sapwood at its periphery. If it is unknown from which part of the tree the test sample was obtained, then the age determined can be substantially in error. The magnitude of such an error may, or may not, be significant, depending on circumstances.

It is hoped this brief note will illustrate some of the reservations regarding the accuracy of radiocarbon dating, apart from those relating to laboratory method and procedure.

Appendix II.
Earthquake Magnitude and Intensity

The following explanation of (a) The Richter Magnitude Scale, and (b) The Modified Mercalli Scale of Felt Intensity, is based on G.A. Eiby (1967).

(a) The Richter Magnitude Scale

The *magnitude* of an earthquake is directly related to the release of seismic energy. The scale was conceived by Charles Richter of the California Institute of Technology, in association with Beno Gutenberg, and introduced in 1935. The original scale was later amended in an attempt to develop a 'unified scale', and this has been used by the U.S. Coast and Geodetic Survey with, regrettably, some consequent confusion. The difference between the scales is not serious, except at low values.

Seismological science differentiates between the magnitude (M) of the various seismic waves received by seismological stations, by the addition of a subscript denoting the type of wave involved. Though the symbol M_L, representing 'local' magnitude of an event, may be encountered elsewhere, for simplicity all magnitudes in this text are represented by (M) alone. The following table gives the approximate equivalence of earthquake magnitude, and energy released by quantities of T.N.T. explosive.

M	T.N.T.	Observations
0	0.25 grams	Equivalent to a man weighing 75 kg jumping off a table
1	15 grams	
2	1 kilo	
2.5		Smallest felt shock
3	65 kilos	
4	3 tonnes	
5	150 tonnes	Often not noticed in earthquake prone regions
5.3		Smallest damaging shocks
6	6 kilotonnes	
6.5	20 kilotonnes	Atomic bomb
7	250 kilotonnes	
8	10 megatonnes	
9	300 megatonnes	

(b) The Modified Mercalli Scale of Felt Intensity

The following is a simplified version for guidance purposes. The *intensity* of a seismic event is always recorded in Roman numerals.

I. Not felt except by few people in exceptional circumstances.

II. Felt only by people at rest, especially on upper floors of buildings. Suspended objects may swing.

III. Felt quite noticeably indoors, but often not recognised as an earthquake. Stationary vehicles may rock, as if from passing vehicles.

IV. Felt indoors by many, outdoors by a few. Dishes, windows, doors disturbed. Walls may crack slightly. Sensation like that of a truck striking a building. Stationary cars rocked.

V. Felt by almost everyone; many awakened. Some dishes, windows and brittle objects may break. Plaster cracked. Trees, poles, and upright objects may sway.

VI. Felt by everyone; many frightened and run outdoors. Heavy furniture moved. Plaster falls and chimneys damaged.

VII. Everyone runs outdoors. Damage slight in well-designed buildings of good construction; slight to moderate in well-built ordinary structures; considerable in poorly built structures; some chimneys broken or toppled.

VIII. Damage slight in specially designed structures; considerable in ordinary substantial buildings with partial collapse; great in poorly built structures. Fall of chimneys, factory stacks, columns, monuments and walls. Heavy furniture overturned. Sand and mud ejected in small amounts. Changes in well water.

IX. Damage considerable, even in well-designed structures. Partial collapse of substantial buildings. Buildings shifted off foundations. Ground cracked conspicuously. Underground pipes broken.

X. Some well-built wooden structures destroyed. Most masonry and frame structures destroyed. Ground badly cracked. Rails bent. Landslides on river banks and hillsides. Water slopped out of ponds, lakes and reservoirs.

XI. Few masonry structures left standing. Bridges destroyed. Broad fissures in ground. Underground pipes severed. Earth slumps and landslides. Rails distorted.

XII. Damage total.

Index

Page numbers in *italic* refer to figures, in **bold** refer to tables and those followed by the letter 'g' refer to glossary entries.

Printed and bound by CPI Group (UK) Ltd, Croydon, CR0 4YY
16/09/2026
14963445-0003